SOLUTIONS MANUAL

SYSTEM DYNAMICS

Fourth Edition

Katsuhiko Ogata

Upper Saddle River, New Jersey 07458

Acquisitions Editor: *Laura Fischer*
Supplement Cover Manager: *Daniel Sandin*
Executive Managing Editor: *Vince O'Brien*
Managing Editor: *David A. George*
Production Editor: *Barbara A. Till*
Supplement Editor: *Andrea Messineo*
Manufacturing Buyer: *Ilene Kahn*

Pearson Prentice Hall
Pearson Education, Inc.
Upper Saddle River, NJ 07458

Printed in the United States of America
10 9 8 7 6 5 4 3 2 1

ISBN 0-13-142463-7

Pearson Education Ltd., *London*
Pearson Education Australia Pty. Ltd., *Sydney*
Pearson Education Singapore, Pte. Ltd.
Pearson Education North Asia Ltd., *Hong Kong*
Pearson Education Canada, Inc., *Toronto*
Pearson Educación de Mexico, S.A. de C.V.
Pearson Education—Japan, *Tokyo*
Pearson Education Malaysia, Pte. Ltd.
Pearson Education, Inc., *Upper Saddle River, New Jersey*

Contents

Preface

This Solutions Manual presents solutions to all unsolved B-problems. For some problems, solutions include more materials than are required in problem statements to aid the user of the book.

The text may be used in a few different ways depending on the course objective and the time allocated to the course.

Sample course coverages are listed below.

If this book is used as a text for a quarter-length course (with approximately 30 lecture hours and 18 recitation hours), Chapters 1 through 7 may be covered.

If the book is used as a text for a semester-length course (with approximately 40 lecture hours and 26 recitation hours), then the first nine chapters may be covered or, alternatively, the first seven chapters plus Chapters 10 and 11 may be covered.

If the course devotes 50 to 60 hours to lectures, then the entire book may be covered in a semester.

The instructor will always have an option to omit certain subjects depending on the course objective.

Katsuhiko Ogata

Solutions to B Problems

CHAPTER 2

B-2-1.

$$f(t) = 0 \qquad t < 0$$
$$= t\,e^{-2t} \qquad t \geq 0$$

Note that

$$\mathcal{L}[t] = \frac{1}{s^2}$$

Referring to Equation (2-2), we obtain

$$F(s) = \mathcal{L}[f(t)] = \mathcal{L}[t\,e^{-2t}] = \frac{1}{(s+2)^2}$$

B-2-2.

(a)

$$f_1(t) = 0 \qquad t < 0$$
$$= 3\sin(5t + 45°) \qquad t \geq 0$$

Note that

$$3\sin(5t + 45°) = 3\sin 5t \cos 45° + 3\cos 5t \sin 45°$$
$$= \frac{3}{\sqrt{2}}\sin 5t + \frac{3}{\sqrt{2}}\cos 5t$$

So we have

$$F_1(s) = \mathcal{L}[f_1(t)] = \frac{3}{\sqrt{2}}\frac{5}{s^2+5^2} + \frac{3}{\sqrt{2}}\frac{s}{s^2+5^2}$$
$$= \frac{3}{\sqrt{2}}\frac{s+5}{s^2+25}$$

(b)

$$f_2(t) = 0 \qquad t < 0$$
$$= 0.03(1 - \cos 2t) \qquad t \geq 0$$

$$F_2(s) = \mathcal{L}[f_2(t)] = 0.03\frac{1}{s} - 0.03\frac{s}{s^2+2^2} = \frac{0.12}{s(s^2+4)}$$

B-2-3.

$$f(t) = 0 \qquad t < 0$$
$$= t^2 e^{-at} \qquad t \geq 0$$

Note that

$$\mathcal{L}[t^2] = \frac{2}{s^3}$$

Referring to Equation (2-2), we obtain

$$F(s) = \mathcal{L}[f(t)] = \mathcal{L}[t^2 e^{-at}] = \frac{2}{(s+a)^3}$$

B-2-4.

$$f(t) = 0 \qquad t < 0$$
$$= \cos 2\omega t \cos 3\omega t \qquad t \geq 0$$

Noting that

$$\cos 2\omega t \cos 3\omega t = \tfrac{1}{2}(\cos 5\omega t + \cos \omega t)$$

we have

$$F(s) = \mathcal{L}[f(t)] = \mathcal{L}[\tfrac{1}{2}(\cos 5\omega t + \cos \omega t)]$$
$$= \frac{1}{2}\left(\frac{s}{s^2 + 25\omega^2} + \frac{s}{s^2 + \omega^2}\right) = \frac{(s^2 + 13\omega^2)s}{(s^2 + 25\omega^2)(s^2 + \omega^2)}$$

B-2-5. The function f(t) can be written as

$$f(t) = (t - a)\, 1(t - a)$$

The Laplace transform of f(t) is

$$F(s) = \mathcal{L}[f(t)] = \mathcal{L}[(t - a)\, 1(t - a)] = \frac{e^{-as}}{s^2}$$

B-2-6.

$$f(t) = c\, 1(t - a) - c\, 1(t - b)$$

The Laplace transform of f(t) is

$$F(s) = c\,\frac{e^{-as}}{s} - c\,\frac{e^{-bs}}{s} = \frac{c}{s}(e^{-as} - e^{-bs})$$

B-2-7. The function f(t) can be written as

$$f(t) = \frac{10}{a^2} - \frac{12.5}{a^2} 1(t - \frac{a}{5}) + \frac{2.5}{a^2} 1(t - a)$$

So the Laplace transform of f(t) becomes

$$F(s) = \mathcal{L}[f(t)] = \frac{10}{a^2}\frac{1}{s} - \frac{12.5}{a^2}\frac{1}{s} e^{-(a/5)s} + \frac{2.5}{a^2}\frac{1}{s} e^{-as}$$

$$= \frac{1}{a^2 s}(10 - 12.5\, e^{-(a/5)s} + 2.5\, e^{-as})$$

As a approaches zero, the limiting value of F(s) becomes as follows:

$$\lim_{a\to 0} F(s) = \lim_{a\to 0} \frac{10 - 12.5\, e^{-(a/5)s} + 2.5\, e^{-as}}{a^2 s}$$

$$= \lim_{a\to 0} \frac{\frac{d}{da}(10 - 12.5\, e^{-(a/5)s} + 2.5\, e^{-as})}{\frac{d}{da} a^2 s}$$

$$= \lim_{a\to 0} \frac{2.5\, s\, e^{-(a/5)s} - 2.5\, s\, e^{-as}}{2as}$$

$$= \lim_{a\to 0} \frac{\frac{d}{da}(2.5\, e^{-(a/5)s} - 2.5\, e^{-as})}{\frac{d}{da} 2a}$$

$$= \lim_{a\to 0} \frac{-0.5\, s\, e^{-(a/5)s} + 2.5\, s\, e^{-as}}{2}$$

$$= \frac{-0.5\, s + 2.5\, s}{2} = \frac{2s}{2} = s$$

B-2-8. The function f(t) can be written as

$$f(t) = \frac{24}{a^3} t - \frac{24}{a^2} 1(t - \frac{a}{2}) - \frac{24}{a^3}(t - a)\, 1(t - a)$$

So the Laplace transform of f(t) becomes

$$F(s) = \frac{24}{a^3}\frac{1}{s^2} - \frac{24}{a^2}\frac{1}{s} e^{-\frac{1}{2}as} - \frac{24}{a^3}\frac{e^{-as}}{s^2}$$

$$= \frac{24}{a^3}\left(\frac{1}{s^2} - \frac{a}{s} e^{-\frac{1}{2}as} - \frac{e^{-as}}{s^2}\right)$$

The limiting value of F(s) as a approaches zero is

$$\lim_{a\to 0} F(s) = \lim_{a\to 0} \frac{24(1 - as\ e^{-\frac{1}{2}as} - e^{-as})}{a^3 s^2}$$

$$= \lim_{a\to 0} \frac{\frac{d}{da}\, 24(1 - as\ e^{-\frac{1}{2}as} - e^{-as})}{\frac{d}{da}\, a^3 s^2}$$

$$= \lim_{a\to 0} \frac{24(-s\ e^{-\frac{1}{2}as} + \frac{as^2}{2}\ e^{-\frac{1}{2}as} + s\ e^{-as})}{3a^2 s^2}$$

$$= \lim_{a\to 0} \frac{\frac{d}{da}\, 8(-\ e^{-\frac{1}{2}as} + \frac{as}{2}\ e^{-\frac{1}{2}as} + e^{-as})}{\frac{d}{da}\, a^2 s}$$

$$= \lim_{a\to 0} \frac{8\left[\frac{s}{2}\ e^{-\frac{1}{2}as} + \frac{s}{2}\ e^{-\frac{1}{2}as} + \frac{as}{2}\left(\frac{-s}{2}\right) e^{-\frac{1}{2}as} - se^{-as}\right]}{2as}$$

$$= \lim_{a\to 0} \frac{\frac{d}{da}\,(4\ e^{-\frac{1}{2}as} - as\ e^{-\frac{1}{2}as} - 4\ e^{-as})}{\frac{d}{da}\, a}$$

$$= \lim_{a\to 0} \frac{-2s\ e^{-\frac{1}{2}as} - s\ e^{-\frac{1}{2}as} + as\ \frac{s}{2}\ e^{-\frac{1}{2}as} + 4s\ e^{-as}}{1}$$

$$= -2s - s + 4s = s$$

B-2-9.

$$f(\infty) = \lim_{t\to\infty} f(t) = \lim_{s\to 0} sF(s)$$

$$= \lim_{s\to 0} \frac{s\ 5(s + 2)}{s(s + 1)} = \frac{5 \times 2}{1} = 10$$

B-2-10.

$$f(0+) = \lim_{t\to 0+} f(t) = \lim_{s\to\infty} \frac{s\ 2(s + 2)}{s(s + 1)(s + 3)} = 0$$

B-2-11. Define

$$y = \dot{x}$$

Then

$$y(0+) = \dot{x}(0+)$$

The initial value of y can be obtained by use of the initial value theorem as follows:

$$y(0+) = \lim_{s \to \infty} sY(s)$$

Since

$$Y(s) = \mathcal{L}_+[y(t)] = \mathcal{L}_+[\dot{x}(t)] = sX(s) - x(0+)$$

we obtain

$$y(0+) = \lim_{s \to \infty} sY(s) = \lim_{s \to \infty} s[sX(s) - x(0+)]$$

$$= \lim_{s \to \infty} [s^2X(s) - sx(0+)]$$

B-2-12. Note that

$$\mathcal{L}\left[\frac{d}{dt} f(t)\right] = sF(s) - f(0)$$

$$\mathcal{L}\left[\frac{d^2}{dt^2} f(t)\right] = s^2F(s) - sf(0) - \dot{f}(0)$$

Define

$$g(t) = \frac{d^2}{dt^2} f(t)$$

Then

$$\mathcal{L}\left[\frac{d^3}{dt^3} f(t)\right] = \mathcal{L}\left[\frac{d}{dt} g(t)\right] = sG(s) - g(0)$$

$$= s[s^2F(s) - sf(0) - \dot{f}(0)] - \ddot{f}(0)$$

$$= s^3F(s) - s^2f(0) - s\dot{f}(0) - \ddot{f}(0)$$

B-2-13.

(a)

$$F_1(s) = \frac{s + 5}{(s + 1)(s + 3)} = \frac{a_1}{s + 1} + \frac{a_2}{s + 3}$$

where

$$a_1 = \left.\frac{s + 5}{s + 3}\right|_{s = -1} = \frac{4}{2} = 2$$

$$a_2 = \left.\frac{s + 5}{s + 1}\right|_{s = -3} = \frac{2}{-2} = -1$$

$F_1(s)$ can thus be written as

$$F_1(s) = \frac{2}{s+1} - \frac{1}{s+3}$$

and the inverse Laplace transform of $F_1(s)$ is

$$f_1(t) = 2\,e^{-t} - e^{-3t}$$

(b)

$$F_2(s) = \frac{3(s+4)}{s(s+1)(s+2)} = \frac{a_1}{s} + \frac{a_2}{s+1} + \frac{a_3}{s+2}$$

where

$$a_1 = \left.\frac{3(s+4)}{(s+1)(s+2)}\right|_{s=0} = \frac{3 \times 4}{2} = 6$$

$$a_2 = \left.\frac{3(s+4)}{s(s+2)}\right|_{s=-1} = \frac{3 \times 3}{(-1) \times 1} = -9$$

$$a_3 = \left.\frac{3(s+4)}{s(s+1)}\right|_{s=-2} = \frac{3 \times 2}{(-2)(-1)} = 3$$

$F_2(s)$ can thus be written as

$$F_2(s) = \frac{6}{s} - \frac{9}{s+1} + \frac{3}{s+2}$$

and the inverse Laplace transform of $F_2(s)$ is

$$f_2(t) = 6 - 9\,e^{-t} + 3\,e^{-2t}$$

B-2-14.

(a)

$$F_1(s) = \frac{6s+3}{s^2} = \frac{6}{s} + \frac{3}{s^2}$$

The inverse Laplace transform of $F_1(s)$ is

$$f_1(t) = 6 + 3t$$

(b)

$$F_2(s) = \frac{5s + 2}{(s + 1)(s + 2)^2} = \frac{a}{s + 1} + \frac{b_2}{(s + 2)^2} + \frac{b_1}{s + 2}$$

where

$$a = \left. \frac{5s + 2}{(s + 2)^2} \right|_{s = -1} = \frac{-5 + 2}{1^2} = -3$$

$$b_2 = \left. \frac{5s + 2}{s + 1} \right|_{s = -2} = \frac{-10 + 2}{-2 + 1} = 8$$

$$b_1 = \left. \frac{d}{ds} \left(\frac{5s + 2}{s + 1} \right) \right|_{s = -2} = \left. \frac{5(s + 1) - (5s + 2)}{(s + 1)^2} \right|_{s = -2}$$

$$= \frac{5(-1) - (-10 + 2)}{1^2} = 3$$

$F_2(s)$ can thus be written as

$$F_2(s) = \frac{-3}{s + 1} + \frac{8}{(s + 2)^2} + \frac{3}{s + 2}$$

and the inverse Laplace transform of $F_2(s)$ is

$$f_2(t) = -3\ e^{-t} + 8t\ e^{-2t} + 3\ e^{-2t}$$

B-2-15.

$$F(s) = \frac{2s^2 + 4s + 5}{s(s + 1)} = 2 + \frac{2}{s + 1} + \frac{5}{s(s + 1)}$$

$$= 2 + \frac{2}{s + 1} + \frac{5}{s} - \frac{5}{s + 1} = 2 - \frac{3}{s + 1} + \frac{5}{s}$$

The inverse Laplace transform of F(s) is

$$f(t) = 2\ \delta(t) - 3\ e^{-t} + 5$$

B-2-16.

$$F(s) = \frac{s^2 + 2s + 4}{s^2} = 1 + \frac{2}{s} + \frac{4}{s^2}$$

The inverse Laplace transform of F(s) is

$$f(t) = \delta(t) + 2 + 4t$$

B-2-17.

$$F(s) = \frac{s}{s^2 + 2s + 10} = \frac{s + 1 - 1}{(s + 1)^2 + 3^2}$$

$$= \frac{s + 1}{(s + 1)^2 + 3^2} - \frac{3}{(s + 1)^2 + 3^2} \frac{1}{3}$$

Hence

$$f(t) = e^{-t} \cos 3t - \frac{1}{3} e^{-t} \sin 3t$$

B-2-18.

$$F(s) = \frac{s^2 + 2s + 5}{s^2(s + 1)} = \frac{a}{s^2} + \frac{b}{s} + \frac{c}{s + 1}$$

where

$$a = \left.\frac{s^2 + 2s + 5}{s + 1}\right|_{s = 0} = 5$$

$$b = \left.\frac{(2s + 2)(s + 1) - (s^2 + 2s + 5)}{(s + 1)^2}\right|_{s = 0} = \frac{2 - 5}{1} = -3$$

$$c = \left.\frac{s^2 + 2s + 5}{s^2}\right|_{s = -1} = \frac{1 - 2 + 5}{1} = 4$$

Hence

$$F(s) = \frac{5}{s^2} + \frac{-3}{s} + \frac{4}{s + 1}$$

The inverse Laplace transform of F(s) is

$$f(t) = 5t - 3 + 4 e^{-t}$$

B-2-19.

$$F(s) = \frac{2s + 10}{(s + 1)^2(s + 4)} = \frac{a}{(s + 1)^2} + \frac{b}{s + 1} + \frac{c}{s + 4}$$

where

$$a = \left.\frac{2s + 10}{s + 4}\right|_{s = -1} = \frac{-2 + 10}{3} = \frac{8}{3}$$

$$b = \left.\frac{2(s + 4) - (2s + 10)}{(s + 4)^2}\right|_{s = -1} = \frac{6 - 8}{3^2} = \frac{-2}{9}$$

$$c = \left.\frac{2s + 10}{(s + 1)^2}\right|_{s = -4} = \frac{-8 + 10}{9} = \frac{2}{9}$$

Hence

$$F(s) = \frac{8}{3}\,\frac{1}{(s + 1)^2} - \frac{2}{9}\,\frac{1}{s + 1} + \frac{2}{9}\,\frac{1}{s + 4}$$

The inverse Laplace transform of F(s) is

$$f(t) = \frac{8}{3}\,te^{-t} - \frac{2}{9}\,e^{-t} + \frac{2}{9}\,e^{-4t}$$

B-2-20.

$$F(s) = \frac{1}{s^2(s^2 + \omega^2)} = \left(\frac{1}{s^2} - \frac{1}{s^2 + \omega^2}\right)\frac{1}{\omega^2}$$

The inverse Laplace transform of F(s) is

$$f(t) = \frac{1}{\omega^2}\left(t - \frac{1}{\omega}\sin \omega t\right)$$

B-2-21.

$$F(s) = \frac{c}{s^2}(1 - e^{-as}) - \frac{b}{s}e^{-as} \qquad a > 0$$

The inverse Laplace transform of F(s) is

$$f(t) = ct - c(t - a)1(t - a) - b\,1(t - a)$$

B-2-22. $\ddot{x} + 4x = 0, \quad x(0) = 5, \quad \dot{x}(0) = 0$

The Laplace transform of the given differential equation is

$$[s^2X(s) - sx(0) - \dot{x}(0)] + 4X(s) = 0$$

Substitution of the initial conditions into this last equation gives

$$(s^2 + 4)X(s) = 5s$$

Solving for X(s), we obtain

$$X(s) = \frac{5s}{s^2 + 4}$$

The inverse Laplace transform of X(s) is

$$x(t) = 5 \cos 2t$$

This is the solution of the given differential equation.

B-2-23. $\ddot{x} + \omega_n^2 x = t, \quad x(0) = 0, \quad \dot{x}(0) = 0$

The Laplace transform of this differential equation is

$$s^2X(s) + \omega_n^2 X(s) = \frac{1}{s^2}$$

Solving this equation for X(s), we obtain

$$X(s) = \frac{1}{s^2(s^2 + \omega_n^2)} = \left(\frac{1}{s^2} - \frac{1}{s^2 + \omega_n^2}\right)\frac{1}{\omega_n^2}$$

The inverse Laplace transform of X(s) is

$$x(t) = \frac{1}{\omega_n^2}\left(t - \frac{1}{\omega_n} \sin \omega_n t\right)$$

This is the solution of the given differential equation.

B-2-24. $2\ddot{x} + 2\dot{x} + x = 1, \quad x(0) = 0, \quad \dot{x}(0) = 2$

The Laplace transform of this differential equation is

$$2[s^2X(s) - sx(0) - \dot{x}(0)] + 2[sX(s) - x(0)] + X(s) = \frac{1}{s}$$

Substitution of the initial conditions into this equation gives

$$2[s^2X(s) - 2] + 2[sX(s)] + X(s) = \frac{1}{s}$$

or

$$(2s^2 + 2s + 1)X(s) = 4 + \frac{1}{s}$$

Solving this last equation for X(s), we get

$$X(s) = \frac{4s + 1}{s(2s^2 + 2s + 1)}$$

$$= \frac{4}{2s^2 + 2s + 1} + \frac{1}{s(2s^2 + 2s + 1)}$$

$$= \frac{2}{(s + 0.5)^2 + 0.25} + \frac{0.5}{s[(s + 0.5)^2 + 0.25]}$$

$$= \frac{4 \times 0.5}{(s + 0.5)^2 + 0.5^2} + \frac{1}{s} - \frac{(s + 0.5) + 0.5}{(s + 0.5)^2 + 0.5^2}$$

The inverse Laplace transform of X(s) gives

$$x(t) = 4\, e^{-0.5t} \sin 0.5t + 1 - e^{-0.5t} \cos 0.5t - e^{-0.5t} \sin 0.5t$$

$$= 1 + 3\, e^{-0.5t} \sin 0.5t - e^{-0.5t} \cos 0.5t$$

B-2-25. $\ddot{x} + x = \sin 3t, \quad x(0) = 0, \quad \dot{x}(0) = 0$

The Laplace transform of this differential equation is

$$s^2X(s) + X(s) = \frac{3}{s^2 + 3^2}$$

Solving this equation for X(s), we get

$$X(s) = \frac{3}{(s^2 + 1)(s^2 + 9)} = \frac{3}{8}\,\frac{1}{s^2 + 1} - \frac{1}{8}\,\frac{3}{s^2 + 9}$$

The inverse Laplace transform of X(s) gives

$$x(t) = \frac{3}{8} \sin t - \frac{1}{8} \sin 3t$$

CHAPTER 3

B-3-1. $J = \frac{1}{2} mR^2 = \frac{1}{2} \times 100 \times 0.5^2 = 12.5 \text{ kg-m}^2$

B-3-2. Assume that the body of known moment of inertia J_0 is turned through a small angle θ about the vertical axis and then released. The equation of motion for the oscillation is

$$J_0 \ddot{\theta} = - k\theta$$

where k is the torsional spring constant of the string. This equation can be written as

$$\ddot{\theta} + \frac{k}{J_0} \theta = 0$$

or

$$\ddot{\theta} + \omega_n^2 \theta = 0$$

where

$$\omega_n = \sqrt{\frac{k}{J_0}}$$

The period T_0 of this oscillation is

$$T_0 = \frac{2\pi}{\omega_n} = \frac{2\pi}{\sqrt{\frac{k}{J_0}}} \qquad (1)$$

Next, we attach a rotating body of unknown moment of inertia J and measure the period T of oscillation. The equation for the period T is

$$T = \frac{2\pi}{\sqrt{\frac{k}{J}}} \qquad (2)$$

By eliminating the unknown torsional spring constant k from Equations (1) and (2), we obtain

$$\frac{2\pi\sqrt{J_0}}{T_0} = \frac{2\pi\sqrt{J}}{T}$$

Hence

$$J = J_0\left(\frac{T}{T_0}\right)^2 \qquad (3)$$

The unknown moment of inertia J can therefore be determined by measuring the period of oscillation T and substituting it into Equation (3).

B-3-3. Define the vertical displacement of the ball as $x(t)$ with $x(0) = 0$. The positive direction is downward. The equation of motion for the system is

$$m\ddot{x} = mg$$

with initial conditions $x(0) = 0$ m and $\dot{x}(0) = 0$ m/s. So we have

$$\ddot{x} = g$$

$$\dot{x} = gt + \dot{x}(0) = gt$$

$$x = \tfrac{1}{2} gt^2 + x(0) = \tfrac{1}{2} gt^2$$

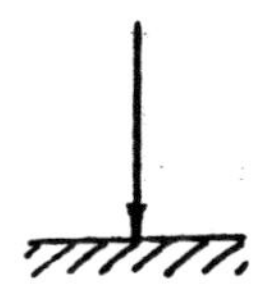

Assume that at $t = t_1$ the ball reaches the ground. Then

$$100 = \tfrac{1}{2} \times 9.807\, t_1^2$$

from which we obtain

$$t_1 = 4.516 \text{ s}$$

The ball reaches the ground in 4.516 s.

The velocity of the ball when it hits the ground is

$$\dot{x}(4.516) = 9.807 \times 4.516 = 44.288 \text{ m/s}$$

B-3-4. Define the torque applied to the flywheel as T. The equation of motion for the system is

$$J\ddot{\theta} = T, \qquad \theta(0) = 0, \qquad \dot{\theta}(0) = 0$$

from which we obtain

$$\dot{\theta} = \frac{T}{J} t$$

By substituting numerical values into this equation, we have

$$20 \times 6.28 = \frac{T}{50} \times 5$$

Thus

$$T = 1256 \text{ N-m}$$

B-3-5. $$J\ddot{\theta} = -T \qquad (T = \text{braking torque})$$

Integrating this equation,

$$\dot{\theta} = -\frac{T}{J} t + \dot{\theta}(0), \qquad \dot{\theta}(0) = 100 \text{ rad/s}$$

Substituting the given numerical values,

$$20 = -\frac{T}{J} \times 15 + 100$$

Solving for T/J, we obtain

$$\frac{T}{J} = 5.33$$

Hence, the deceleration given by the brake is 5.33 rad/s^2.

The total angle rotated in 15-second period is obtained from

$$\theta(t) = -\frac{T}{J}\frac{t^2}{2} + \dot{\theta}(0)t + \theta(0), \qquad \theta(0) = 0, \qquad \dot{\theta}(0) = 100$$

as follows;

$$\theta(15) = -5.33 \times \frac{15^2}{2} + 100 \times 15 = 900 \text{ rad}$$

B-3-6. The equations for the system are

$$F = k_2(x - y)$$

$$k_2(x - y) = k_1 y$$

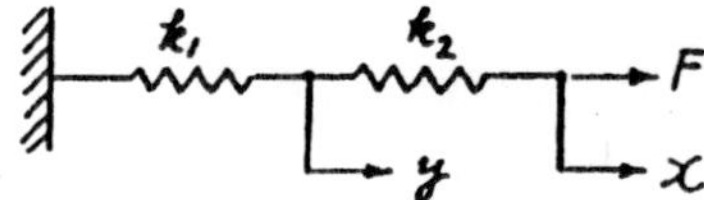

Eliminating y from the two equations gives

$$F = k_2\left(x - \frac{k_2 x}{k_1 + k_2}\right) = \frac{k_1 k_2}{k_1 + k_2} x = \frac{1}{\frac{1}{k_1} + \frac{1}{k_2}} x$$

The equivalent spring constant k_{eq} is then obtained as

$$k_{eq} = \frac{1}{\frac{1}{k_1} + \frac{1}{k_2}}$$

Next, consider the figure shown below. Note that ΔABD and ΔCBE are similar. So we have

$$\frac{\overline{CE}}{\overline{AD}} = \frac{\overline{BE}}{\overline{BD}}$$

which can be rewritten as

$$\frac{\overline{OC}\,\frac{\sqrt{3}}{2}}{\overline{OA}\,\frac{\sqrt{3}}{2}} = \frac{\overline{OB} - \overline{OC}\,\frac{1}{2}}{\overline{OB} + \overline{OA}\,\frac{1}{2}}$$

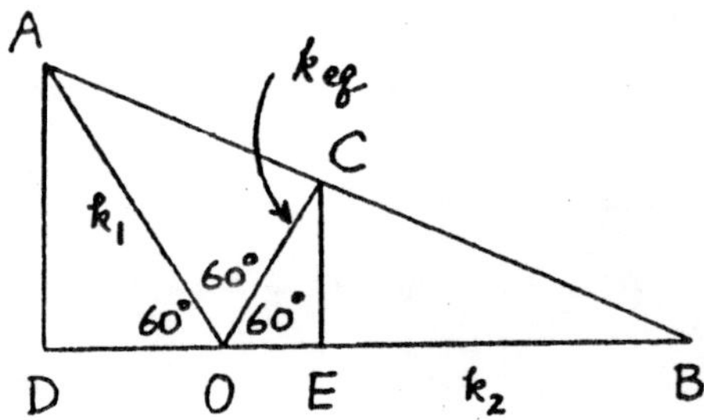

or

$$\overline{OC}(\overline{OB} + \tfrac{1}{2}\,\overline{OA}) = \overline{OA}(\overline{OB} - \tfrac{1}{2}\,\overline{OC})$$

Solving for $\overline{OC}$, we obtain

$$\overline{OC} = \frac{\overline{OA}\ \overline{OB}}{\overline{OA} + \overline{OB}} = \frac{1}{\frac{1}{\overline{OA}} + \frac{1}{\overline{OB}}} = \frac{1}{\frac{1}{k_1} + \frac{1}{k_2}} = k_{eq}$$

B-3-7. Assume that we apply force F to the spring system. Then

$$F = k_1 x + k_2(x - y)$$

$$k_2(x - y) = k_3 y$$

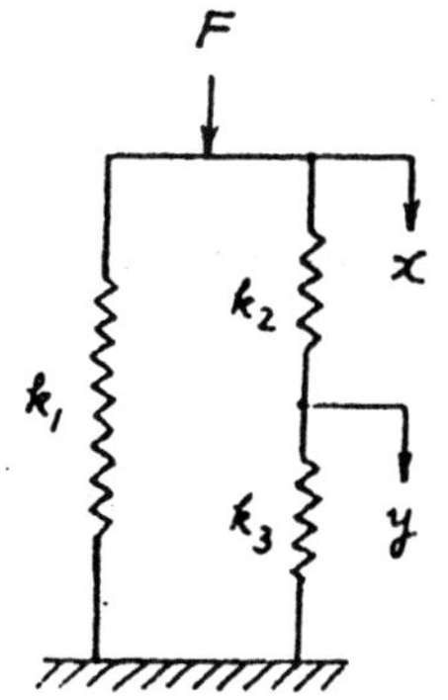

Eliminating y from the preceding equations, we obtain

$$F = \frac{k_1(k_2 + k_3) + k_2 k_3}{k_2 + k_3} x$$

$$= \left(k_1 + \frac{1}{\frac{1}{k_2} + \frac{1}{k_3}} \right) x$$

Hence the equivalent spring constant k_{eq} is given by

$$k_{eq} = k_1 + \frac{1}{\frac{1}{k_2} + \frac{1}{k_3}}$$

B-3-8.

(a) The force f due to the dampers is

$$f = b_1(\dot{y} - \dot{x}) + b_2(\dot{y} - \dot{x}) = (b_1 + b_2)(\dot{y} - \dot{x})$$

In terms of the equivalent viscous friction coefficient b_{eq}, force f is given by

$$f = b_{eq}(\dot{y} - \dot{x})$$

Hence

$$b_{eq} = b_1 + b_2$$

(b) The force f due to the dampers is

$$f = b_1(\dot{z} - \dot{x}) = b_2(\dot{y} - \dot{z}) \qquad (1)$$

where z is the displacement of a point between damper b_1 and damper b_2. (Note that the same force is transmitted through the shaft.) From Equation (1), we have

$$(b_1 + b_2)\dot{z} = b_2\dot{y} + b_1\dot{x}$$

or

$$\dot{z} = \frac{1}{b_1 + b_2}(b_2\dot{y} + b_1\dot{x}) \qquad (2)$$

In terms of the equivalent viscous friction coefficient b_{eq}, force f is given by

$$f = b_{eq}(\dot{y} - \dot{x})$$

By substituting Equation (2) into Equation (1), we have

$$f = b_2(\dot{y} - \dot{z}) = b_2[\dot{y} - \frac{1}{b_1 + b_2}(b_2\dot{y} + b_1\dot{x})]$$

$$= \frac{b_1 b_2}{b_1 + b_2}(\dot{y} - \dot{x})$$

Thus,

$$f = b_{eq}(\dot{y} - \dot{x}) = b_2(\dot{y} - \dot{z}) = \frac{b_1 b_2}{b_1 + b_2}(\dot{y} - \dot{x})$$

Hence,

$$b_{eq} = \frac{b_1 b_2}{b_1 + b_2} = \frac{1}{\frac{1}{b_1} + \frac{1}{b_2}}$$

B-3-9. Since the same force transmits the shaft, we have

$$f = b_1(\dot{z} - \dot{x}) = b_2(\dot{y} - \dot{z}) + b_3(\dot{y} - \dot{z}) \qquad (1)$$

where displacement z is defined in the figure below.

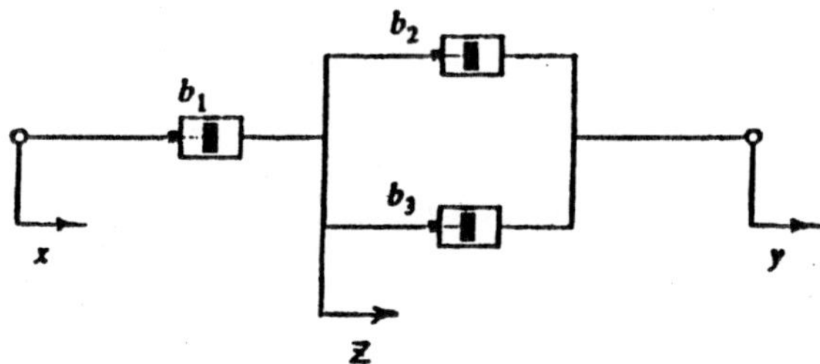

In terms of the equivalent viscous friction coefficient, the force f is given by

$$f = b_{eq}(\dot{y} - \dot{x}) \qquad (2)$$

From Equation (1) we have

$$b_1\dot{z} + b_2\dot{z} + b_3\dot{z} = b_1\dot{x} + b_2\dot{y} + b_3\dot{y}$$

or

$$\dot{z} = \frac{1}{b_1 + b_2 + b_3}[b_1\dot{x} + (b_2 + b_3)\dot{y}] \qquad (3)$$

By substituting Equation (3) into Equation (1), we have

$$f = b_1(\dot{z} - \dot{x}) = b_1 \left\{ \frac{1}{b_1 + b_2 + b_3} [b_1\dot{x} + (b_2 + b_3)\dot{y}] - \dot{x} \right\}$$

$$= b_1 \frac{b_2 + b_3}{b_1 + b_2 + b_3} (\dot{y} - \dot{x}) \qquad (4)$$

Hence, by comparing Equations (2) and (4), we obtain

$$b_{eq} = b_1 \left(\frac{b_2 + b_3}{b_1 + b_2 + b_3} \right) = \frac{1}{\frac{1}{b_2 + b_3} + \frac{1}{b_1}}$$

B-3-10. The equation for the system is

$$m\ddot{x} = -(k_1 + k_2)x - k_3 x$$

or

$$m\ddot{x} + (k_1 + k_2 + k_3)x = 0$$

The natural frequency of the system is

$$\omega_n = \sqrt{\frac{k_1 + k_2 + k_3}{m}}$$

B-3-11. The density ρ of the liquid is

$$\rho = \frac{m}{LA}$$

where A is the cross-sectional area of the inside of the glass tube. The mathematical model for the system is

$$m\ddot{x} = -\rho Ag\, 2x$$

or

$$\ddot{x} + \frac{2g}{L} x = 0$$

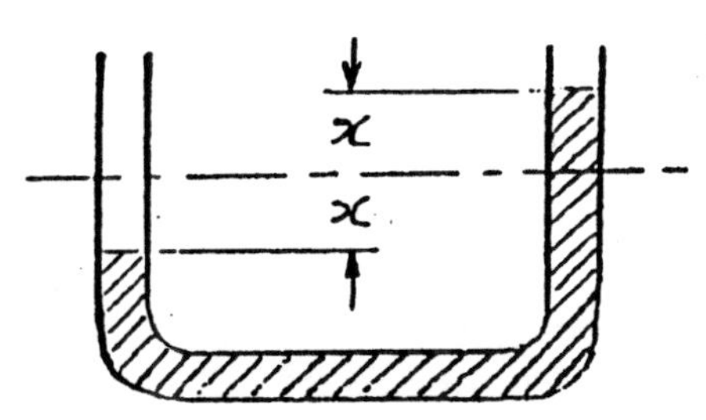

The natural frequency is

$$\omega_n = \sqrt{\frac{2g}{L}}$$

B-3-12. For a small displacement x, the torque balance equation for the system is

$$m\ddot{x}(2a) = -k(\frac{1}{2}x)a$$

or

$$m\ddot{x} + \frac{k}{4}x = 0$$

The natural frequency is

$$\omega_n = \frac{1}{2}\sqrt{\frac{k}{m}} = \frac{1}{2}\sqrt{\frac{400}{\frac{5}{9.81}}} = 14.01 \text{ rad/sec}$$

B-3-13. Applying Newton's second law to the system, we obtain

$$J\ddot{\theta}_o = -b\dot{\theta}_o - k(\theta_o - \theta_i)$$

Hence

$$J\ddot{\theta}_o + b\dot{\theta}_o + k\theta_o = k\theta_i$$

This is a mathematical model for the system.

B-3-14. A mathematical model for the system is

$$m\ddot{x} = -k_1x - b_1\dot{x} - k_2x - b_2\dot{x}$$

or

$$m\ddot{x} + (b_1 + b_2)\dot{x} + (k_1 + k_2)x = 0$$

B-3-15. The equation of motion for the system is

$$m\ddot{x} + b\dot{x} + kx = 0$$

Substituting the given numerical values for m, b, and k into this equation, we obtain

$$2\ddot{x} + 4\dot{x} + 20x = 0 \qquad (1)$$

where $x(0) = 0.1$ and $\dot{x}(0) = 0$. The response to the given initial condition can be obtained by taking the Laplace transform of this equation, solving the resulting equation for $X(s)$, and finding the inverse Laplace transform of $X(s)$. The Laplace transform of Equation (1) is

$$2[s^2X(s) - sx(0) - \dot{x}(0)] + 4[sX(s) - x(0)] + 20\, X(s) = 0$$

By substituting the given initial conditions into this last equation, we get

$$2[s^2X(s) - 0.1s] + 4[sX(s) - 0.1] + 20\ X(s) = 0$$

Solving this equation for X(s) gives

$$X(s) = \frac{0.2s + 0.4}{2s^2 + 4s + 20} = \frac{0.1 + 0.1(s+1)}{(s+1)^2 + 3^2}$$

The inverse Laplace transform of this last equation gives

$$x(t) = 0.1(\frac{1}{3} e^{-t} \sin 3t + e^{-t} \cos 3t)$$

B-3-16. The equations of motion for the system are

$$J\ddot{\theta} = (T_1 - T_2)R$$

$$m\ddot{x} = - T_1$$

$$M\ddot{y} = - ky + T_1 + T_2$$

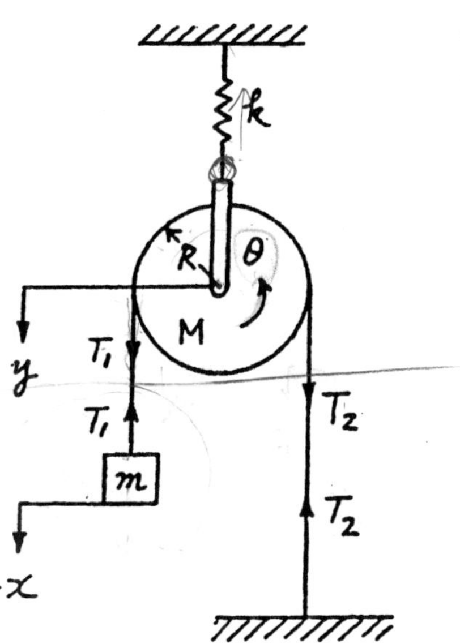

Noting that $x = 2y$, $R\theta = x - y = y$, and $J = \frac{1}{2}MR^2$, the three equations can be rewritten as

$$\frac{1}{2} MR^2\ddot{\theta} = \frac{1}{2} MR\ddot{y} = (T_1 - T_2)R$$

$$m\ddot{x} = - T_1$$

$$M\ddot{y} + ky = T_1 + T_2$$

Eliminating T_2 from the preceding equations gives

$$\frac{1}{2} M\ddot{y} + M\ddot{y} + ky = 2T_1 = - 2m\ddot{x}$$

By changing y into x,

$$\frac{3}{2} M \frac{\ddot{x}}{2} + k \frac{x}{2} = - 2m\ddot{x}$$

or

$$(m + \frac{3}{8} M)\ddot{x} + \frac{1}{4} kx = 0$$

The natural frequency is

$$\omega_n = \sqrt{\frac{2k}{8m + 3M}}$$

If mass m is pulled down a distance x_o and released with zero initial velocity, the motion of mass m is

$$x(t) = x_o \cos \sqrt{\frac{2k}{8m + 3M}}\, t$$

B-3-17. Referring to the figure below, we have

$$m\ddot{x} = -T \qquad (1)$$

where T is the tension in the wire. (Note that since x is measured from the static equilibrium position, the term mg does not enter the equation.) For the rotational motion of the pulley, we have

$$J\ddot{\theta} = -k_2(y + R_2\theta)R_2 + k_2(y - R_2\theta)R_2 + TR_1 - k_1R_1x$$

or

$$J\ddot{\theta} = -k_2R_2^2\theta - k_2R_2^2\theta + TR_1 - k_1R_1x \qquad (2)$$

Eliminating T from Equations (1) and (2), we obtain

$$J\ddot{\theta} + 2k_2R_2^2\theta + mR_1\ddot{x} + k_1R_1x = 0$$

Since $x = R_1\theta$, we have

$$J\ddot{\theta} + 2k_2R_2^2\theta + mR_1^2\ddot{\theta} + k_1R_1^2\theta = 0$$

or

$$(J + mR_1^2)\ddot{\theta} + (2k_2R_2^2 + k_1R_1^2)\theta = 0$$

or

$$\ddot{\theta} + \frac{2k_2R_2^2 + k_1R_1^2}{J + mR_1^2}\theta = 0$$

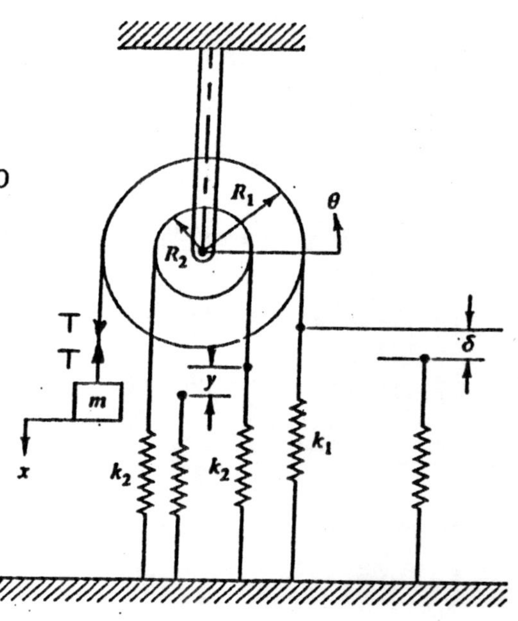

This last equation is a mathematical model of the system. The natural frequency of the system is

$$\omega_n = \sqrt{\frac{2k_2R_2^2 + k_1R_1^2}{J + mR_1^2}}$$

B-3-18. Torque = T = 50 x 0.5 = 25 N-m

Power = $T\omega$ = 25 x 100 = 2500 N-m/s = 2500 W

B-3-19. The potential energy of the system consists of the potential energy of the mass m and that of the springs. Choose the datum line such that the potential energy at the equilibrium position (where the pendulum is vertical) to be zero. Then the potential energy, when the pendulum is vibrating, is

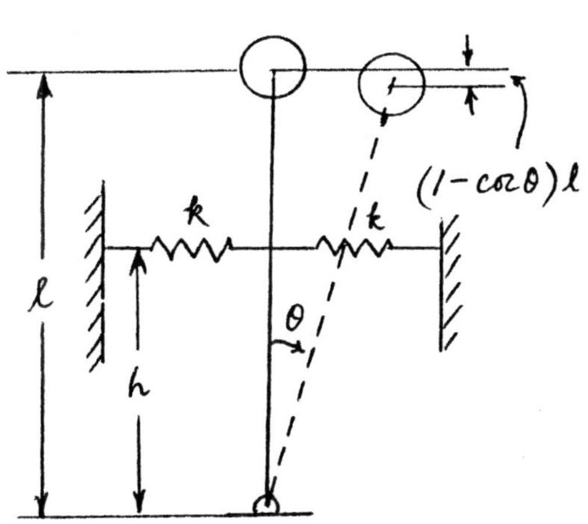

$$U = -(1 - \cos\theta)\ell mg + \tfrac{1}{2}k(\theta h)^2 + \tfrac{1}{2}k(\theta h)^2$$

$$= -(1 - \cos\theta)\ell mg + k\theta^2 h^2$$

Since the expression for U involves θ^2 but does not involve θ, it is necessary to expand $\cos\theta$ into the following infinite series:

$$\cos\theta = 1 - \frac{\theta^2}{2!} + \frac{\theta^4}{4!} - \cdots$$

and retain only the first two terms, or

$$\cos\theta = 1 - \tfrac{1}{2}\theta^2$$

The potential energy U for small θ can then be written as

$$U = -(1 - 1 + \tfrac{1}{2}\theta^2)\ell mg + k\theta^2 h^2$$

$$= (kh^2 - \tfrac{1}{2}\ell mg)\theta^2$$

The kinetic energy T when the pendulum is vibrating is

$$T = \tfrac{1}{2}m(\ell\dot{\theta})^2 = \tfrac{1}{2}m\ell^2\dot{\theta}^2$$

When the pendulum is vibrating, we can assume that

$$\theta = A\sin\omega t$$

where A is the amplitude of vibration. By substituting the preceding expression for θ into the potential energy U and kinetic energy T, we obtain

$$U = (kh^2 - \tfrac{1}{2}\ell mg)A^2\sin^2\omega t$$

$$T = \tfrac{1}{2}m\ell^2 A^2\omega^2\cos^2\omega t$$

Hence, U_{max} and T_{max} can be obtained as

$$U_{max} = (kh^2 - \tfrac{1}{2}\ell mg)A^2, \qquad T_{max} = \tfrac{1}{2}m\ell^2 A^2\omega^2$$

By equating U_{max} and T_{max}, we have

$$(kh^2 - \tfrac{1}{2}\ell mg)A^2 = \tfrac{1}{2}m\ell^2A^2\omega^2$$

which can be simplified to

$$\omega^2 = \frac{2kh^2 - \ell mg}{m\ell^2} = \frac{2kh^2}{m\ell^2} - \frac{g}{\ell}$$

Hence, the frequency of vibration is

$$\omega = \sqrt{\frac{2kh^2}{m\ell^2} - \frac{g}{\ell}}$$

Note that this expression is valid for small θ such that $\cos\theta$ can be approximated by $1 - \tfrac{1}{2}\theta^2$.

<u>B-3-20</u>. The kinetic energy T is

$$T = \frac{1}{2}M\ell^2\dot{\theta}^2 + \frac{1}{2}\int_0^\ell \frac{m}{\ell}\dot{\theta}^2\xi^2\,d\xi = \frac{1}{2}(M + \frac{m}{3})\ell^2\dot{\theta}^2$$

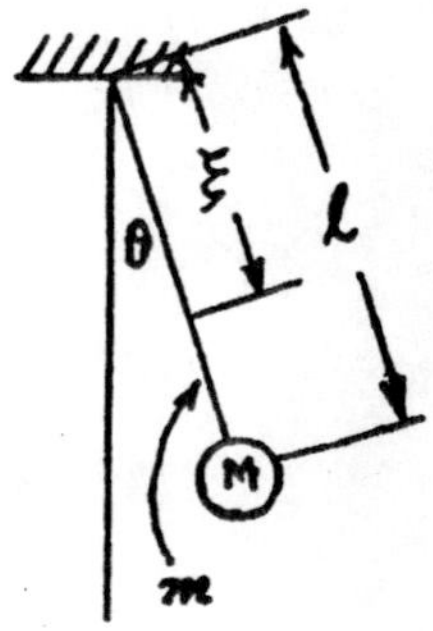

The potential energy U is

$$U = Mg\ell(1 - \cos\theta) + \int_0^\ell \frac{m}{\ell}g\xi(1 - \cos\theta)\,d\xi$$

$$= (M + \frac{m}{2})g\ell(1 - \cos\theta)$$

Since the system is conservative, we have

$$T + U = \frac{1}{2}(M + \frac{m}{3})\ell^2\dot{\theta}^2 + (M + \frac{m}{2})g\ell(1 - \cos\theta)$$

$$= \text{constant}$$

Noting that $d(T + U)/dt = 0$, we obtain

$$(M + \frac{m}{3})\ell^2\dot{\theta}\ddot{\theta} + (M + \frac{m}{2})g\ell\sin\theta\,\dot{\theta} = 0$$

or

$$\left[(M + \frac{m}{3})\ell^2\ddot{\theta} + (M + \frac{m}{2})g\ell\sin\theta\right]\dot{\theta} = 0$$

Since θ is not identically zero, we have

$$(M + \frac{m}{3})\ell^2\ddot{\theta} + (M + \frac{m}{2})g\ell\sin\theta = 0$$

Rewriting,

$$\ddot{\theta} + \frac{M + \frac{m}{2}}{M + \frac{m}{3}} \frac{g}{\ell} \sin \theta = 0$$

For small values of θ,

$$\ddot{\theta} + \frac{M + \frac{m}{2}}{M + \frac{m}{3}} \frac{g}{\ell} \theta = 0$$

So the natural frequency is

$$\omega_n = \sqrt{\frac{M + \frac{m}{2}}{M + \frac{m}{3}} \frac{g}{\ell}}$$

CHAPTER 4

B-4-1. The system equations are

$$k_1(x_i - y) = b_1(\dot{y} - \dot{x}_o)$$

$$b_1(\dot{y} - \dot{x}_o) = k_2 x_o$$

which can be rewritten as

$$b_1\dot{y} + k_1 y = k_1 x_i + b_1\dot{x}_o$$

$$b_1\dot{x}_o + k_2 x_o = b_1\dot{y}$$

Noting that $x_o(0-) = 0$ and $y(0-) = 0$, by taking the $\mathcal{L}_-$ transform of these two equations we obtain

$$(b_1 s + k_1)Y(s) = k_1 X_i(s) + b_1 s X_o(s)$$

$$(b_1 s + k_2)X_o(s) = b_1 s Y(s)$$

By eliminating Y(s) from the preceding two equations, we get

$$(b_1 s + k_1)\frac{b_1 s + k_2}{b_1 s} X_o(s) = k_1 X_i(s) + b_1 s X_o(s)$$

Simplifying,

$$[(k_1 + k_2)b_1 s + k_1 k_2]\, X_o(s) = k_1 b_1 s X_i(s)$$

or

$$\frac{X_o(s)}{X_i(s)} = \frac{k_1 b_1 s}{(k_1 + k_2)b_1 s + k_1 k_2} = \frac{\frac{b_1}{k_2} s}{\left(\frac{1}{k_1} + \frac{1}{k_2}\right) b_1 s + 1}$$

The response of the system to $x_i(t) = X_i 1(t)$ can be obtained by taking the inverse Laplace transform of

$$X_o(s) = \frac{(b_1/k_2)s}{\left(\frac{1}{k_1} + \frac{1}{k_2}\right) b_1 s + 1} \frac{X_i}{s}$$

$$= \frac{k_1 X_i}{k_1 + k_2} \frac{1}{s + \frac{1}{\left(\frac{1}{k_1} + \frac{1}{k_2}\right) b_1}}$$

as follows:

$$x_o(t) = \frac{k_1 X_i}{k_1 + k_2} \exp\left\{-t/[(\frac{1}{k_1} + \frac{1}{k_2})b_1]\right\} \qquad t \geqslant 0$$

<u>B-4-2</u>. The equations of motion for the system are

$$k_1(x_i - x_o) = b_2(\dot{x}_o - \dot{y})$$

$$b_2(\dot{x}_o - \dot{y}) = k_2 y$$

Rewriting these equations,

$$b_2\dot{x}_o + k_1 x_o = k_1 x_i + b_2\dot{y}$$

$$b_2\dot{y} + k_2 y = b_2\dot{x}_o$$

Noting that $x(0-) = 0$ and also $y(0-) = 0$, $\mathcal{L}_-$ transforms of these two equations become

$$(b_2 s + k_1)X_o(s) = k_1 X_i(s) + b_2 sY(s)$$

$$(b_2 s + k_2)Y(s) = b_2 sX_o(s)$$

Eliminating $Y(s)$ from the last two equations, we obtain

$$(b_2 s + k_1)X_o(s) = k_1 X_i(s) + b_2 s \frac{b_2 sX_o(s)}{b_2 s + k_2}$$

which can be simplified as

$$[(k_1 + k_2)b_2 s + k_1 k_2]X_o(s) = k_1(b_2 s + k_2)X_i(s)$$

Hence

$$\frac{X_o(s)}{X_i(s)} = \frac{k_1(b_2 s + k_2)}{(k_1 + k_2)b_2 s + k_1 k_2}$$

Since the input $x_i(t)$ is given as

$$x_i(t) = X_i \qquad 0 < t < t_1$$

$$= 0 \qquad \text{elsewhere}$$

we have

$$X_i(s) = \frac{X_i}{s}(1 - e^{-t_1 s})$$

The response $X_o(s)$ is then obtained as

$$X_o(s) = \frac{k_1(b_2 s + k_2)}{(k_1 + k_2)b_2 s + k_1 k_2} \frac{X_i}{s}(1 - e^{-t_1 s})$$

Since

$$\frac{k_1(b_2 s + k_2)}{(k_1 + k_2)b_2 s + k_1 k_2} \frac{1}{s} = \frac{1}{s} + \frac{-\frac{b_2}{k_1}}{\left(\frac{1}{k_1} + \frac{1}{k_2}\right)b_2 s + 1}$$

we have

$$\mathcal{L}^{-1}\left[\frac{k_1(b_2 s + k_2)}{(k_1 + k_2)b_2 s + k_1 k_2} \frac{1}{s}\right] = 1 - \frac{k_2}{k_1 + k_2} \exp\left\{-t/[(\frac{1}{k_1} + \frac{1}{k_2})b_2]\right\}$$

Hence

$$x_o(t) = X_i\left[1 - \frac{k_2}{k_1 + k_2} \exp\left\{-t/[(\frac{1}{k_1} + \frac{1}{k_2})b_2]\right\}\right]$$

$$- X_i\left[1 - \frac{k_2}{k_1 + k_2} \exp\left\{-(t - t_1)/[(\frac{1}{k_1} + \frac{1}{k_2})b_2]\right\}\right]1(t - t_1)$$

B-4-3. Define the displacement of a point between springs k_1 and k_2 as y. Then, the equations of motion for the system become

$$m\ddot{x} + k_2(x - y) = u$$

$$k_1 y = k_2(x - y)$$

From the second equation, we have

$$k_1 y + k_2 y = k_2 x$$

or

$$y = \frac{k_2}{k_1 + k_2} x$$

Then

$$m\ddot{x} + \frac{k_1 k_2}{k_1 + k_2} x = u$$

or

$$\frac{X(s)}{U(s)} = \frac{1}{ms^2 + \dfrac{k_1 k_2}{k_1 + k_2}}$$

B-4-4. We assume that the system remains linear throughout the response period. Define the displacement of the point where force u is applied (point A in the diagram shown below) as y.

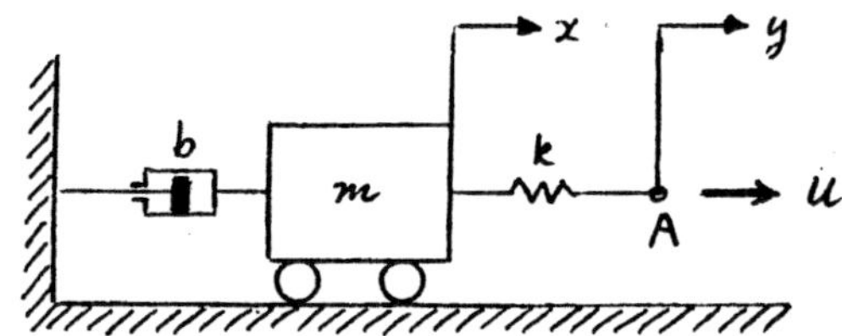

Then, the equations for the system are

$$m\ddot{x} = - b\dot{x} - k(x - y)$$

$$u = k(y - x)$$

Hence

$$m\ddot{x} + b\dot{x} = k(y - x) = u$$

By taking the Laplace transform of this last equation, assuming zero initial conditions, we obtain

$$(ms^2 + bs)X(s) = U(s)$$

Hence

$$\frac{X(s)}{U(s)} = \frac{1}{ms^2 + bs}$$

Note that by applying the force u at point A, the spring is elongated by y - x. The spring exerts the pulling force k(y - x) (which is equal to the force u) to the mass m. As far as the motion of the mass m is concerned, applying the force u through the spring is the same as applying the force u directly to the mass m.)

B-4-5. We assume that the system remains linear during the entire period of response. Define displacements y and z as shown in the figure below. Then, for the given system, the following two equations can be derived:

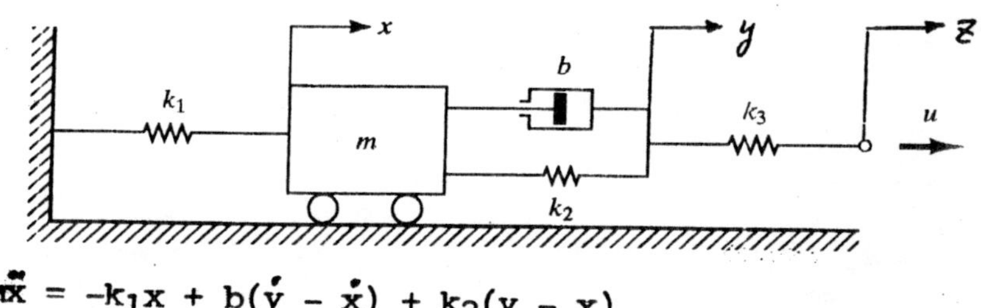

$$m\ddot{x} = -k_1 x + b(\dot{y} - \dot{x}) + k_2(y - x)$$

$$b(\dot{y} - \dot{x}) + k_2(y - x) = k_3(z - y) = u$$

Rewriting, we get

$$m\ddot{x} + b\dot{x} + k_1 x + k_2 x = b\dot{y} + k_2 y$$

$$b\dot{y} + k_2 y = b\dot{x} + k_2 x + u$$

Eliminating $b\dot{y} + k_2 y$ from these two equations, we obtain

$$m\ddot{x} + b\dot{x} + k_1 x + k_2 x = b\dot{x} + k_2 x + u$$

or

$$m\ddot{x} + k_1 x = u$$

Hence the transfer function X(s)/U(s) is obtained as

$$\frac{X(s)}{U(s)} = \frac{1}{ms^2 + k_1}$$

Note that the response of the mass m shown in Figure 4-54 is the same as the response of the mass m of the system shown below.

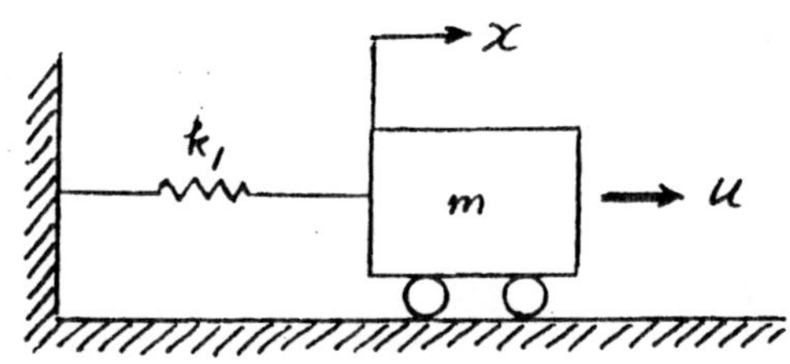

B-4-6. We assume that the system is to remain linear during the entire period of response. Consider the system shown below, where we added the second cart with mass m_0. We derive equations for this system and then let $m_0 \to 0$.

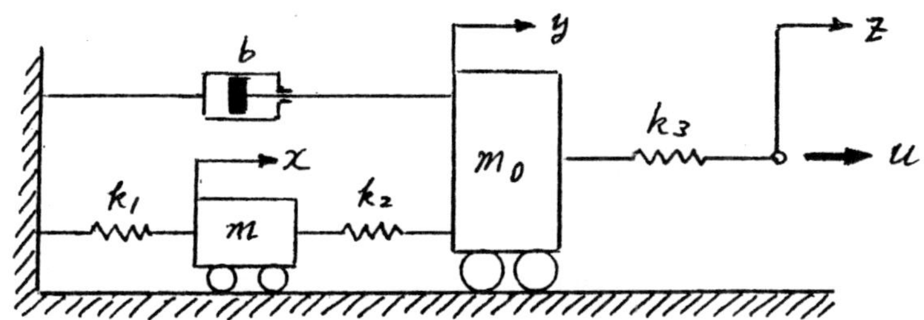

The equations of motion for the system are

$$m\ddot{x} = -k_1 x - k_2(x - y)$$

$$m_0\ddot{y} = -b\dot{y} - k_2(y - x) - k_3(y - z)$$

$$u = k_3(z - y)$$

By letting $m_0 \to 0$ and rewriting these equations, we obtain

$$m\ddot{x} + (k_1 + k_2)x = k_2 y$$

$$b\dot{y} + k_2(y - x) = u$$

Taking the Laplace transforms of these equations, assuming zero initial conditions, we get

$$(ms^2 + k_1 + k_2)X(s) = k_2 Y(s)$$

$$(bs + k_2)Y(s) = k_2 X(s) + U(s)$$

Eliminating Y(s) from the last two equation, we obtain

$$(ms^2 + k_1 + k_2)X(s) = \frac{k_2[k_2 X(s) + U(s)]}{bs + k_2}$$

Simplifying,

$$[mbs^3 + mk_2 s^2 + (k_1 + k_2)bs + k_1 k_2]X(s) = k_2 U(s)$$

from which we get the transfer function X(s)/U(s) as follows:

$$\frac{X(s)}{U(s)} = \frac{k_2}{mbs^3 + mk_2 s^2 + (k_1 + k_2)bs + k_1 k_2}$$

B-4-7. The equation for the system is

$$b(\dot{x} - L\dot{\theta}) = kL\theta$$

or

$$L\dot{\theta} + \frac{k}{b} L\theta = \dot{x}$$

The Laplace transform of this last equation, assuming zero initial conditions, is

$$\left(Ls + \frac{k}{b} L\right) \Theta(s) = sX(s)$$

Hence

$$\frac{\Theta(s)}{X(s)} = \frac{1}{L} \frac{s}{s + \frac{k}{b}}$$

Thus, the system is a differentiating element.

For the unit-step input X(s) = 1/s, the output $\Theta(s)$ becomes

$$\Theta(s) = \frac{1}{L} \frac{1}{s + \frac{k}{b}}$$

The inverse Laplace transform of $\Theta(s)$ gives

$$\theta(t) = \frac{1}{L} e^{-(k/b)t}$$

The response curve $\theta(t)$ versus t is an exponentially decaying curve.

B-4-8. The system equation is

$$m\ddot{x} = -b(\dot{x} - \dot{u}) - k(x - u)$$

Hence

$$m\ddot{x} + b\dot{x} + kx = b\dot{u} + ku$$

Taking the Laplace transform of this equation, assuming zero initial conditions, we obtain

$$(ms^2 + bs + k)X(s) = (bs + k)U(s) \tag{1}$$

Rewriting Equation (1) in the following way,

$$X(s) = \frac{bs + k}{ms^2 + bs + k} U(s)$$

we get the block diagram shown in Figure (a).

Figure (a)

$$U(s) \rightarrow \boxed{\frac{bs+k}{ms^2+bs+k}} \rightarrow X(s)$$

By rewriting Equation (1) in the following way

$$(ms^2 + bs)X(s) = (bs + k)U(s) - kX(s)$$

we obtain the block diagram as shown in Figure (b).

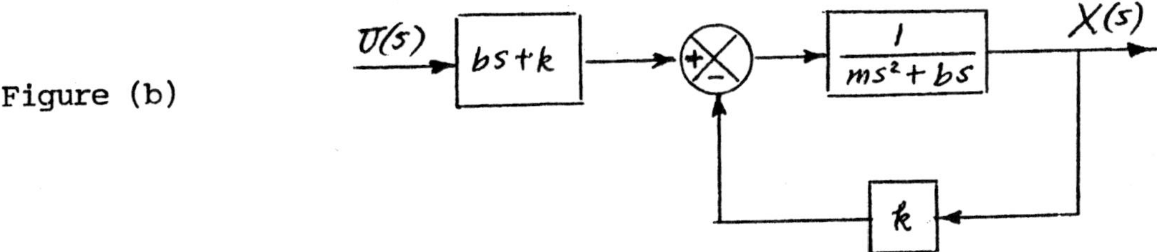

Figure (b)

If we rewrite Equation (1) as

$$ms^2X(s) = (bs + k)[U(s) - X(s)]$$

we get the block diagram as shown in Figure (c).

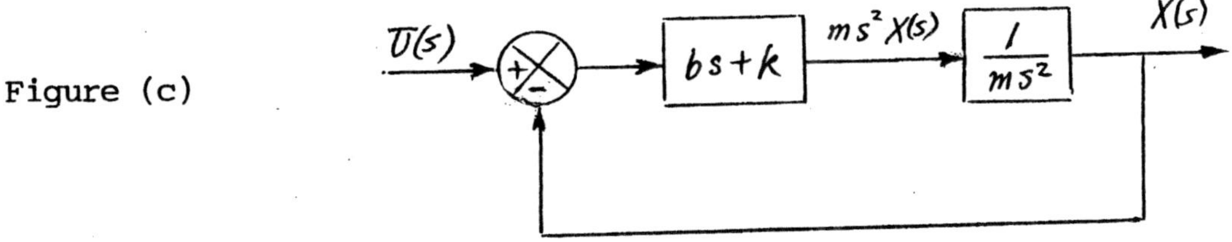

Figure (c)

Also, by rewriting Equation (1) as

$$(bs + k)U(s) - [kX(s) + bsX(s)] = ms^2X(s)$$

and dividing all terms by m, we get

$$\frac{bs + k}{m} U(s) - \frac{k}{m} X(s) - \frac{b}{m} sX(s) = s^2X(s)$$

From this equation we get the block diagram shown in Figure (d).

Figure (d)

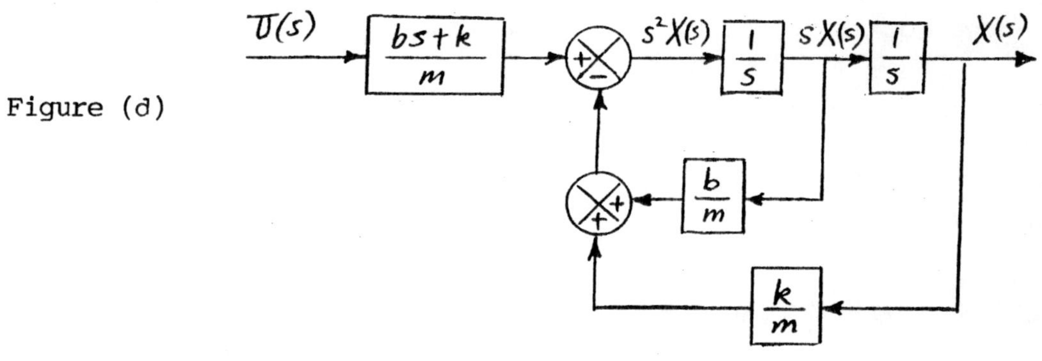

B-4-9. A MATLAB program for obtaining the partial-fraction expansion of the given B(s)/A(s) is shown below.

```
>> num = [1];
>> den = [1    1    81    81    0];
>> format long
>> [r,p,k] = residue(num,den)

r =

  -0.00007527853056 + 0.00067750677507i
  -0.00007527853056 - 0.00067750677507i
  -0.01219512195122
   0.01234567901235


p =

  -0.00000000000000 + 9.00000000000001i
  -0.00000000000000 - 9.00000000000001i
  -1.00000000000000
                  0


k =

     []
```

From the MATLAB output, we get the following expression:

$$\frac{B(s)}{A(s)} = \frac{-0.00007528 + j0.0006775}{s - j9} + \frac{-0.00007528 - j0.0006775}{s + j9}$$

$$+ \frac{-0.0121951}{S + 1} + \frac{0.01234568}{s}$$

$$= -\frac{0.00015056s + 0.012195}{s^2 + 81} - \frac{0.0121951}{s + 1} + \frac{0.01234568}{s}$$

B-4-10. A MATLAB program for obtaining the partial-fraction expansion of the given B(s)/A(s) is given below.

```
>> num = [5    10];
>> den = [1    5    19    12    0    0];
>> [r,p,k] = residue(num,den)

r =

   0.0436 + 0.0042i
   0.0436 - 0.0042i
   0.8157
  -0.9028
   0.8333


p =

  -2.1197 + 3.3589i
  -2.1197 - 3.3589i
  -0.7607
        0
        0


k =

     []
```

From the MATLAB output, we get the following expression:

$$\frac{B(s)}{A(s)} = \frac{0.0436 + j0.0042}{s + 2.1197 - j3.3589} + \frac{0.0436 - j0.0042}{s + 2.1197 + j3.3589}$$

$$+ \frac{0.8157}{s + 0.7607} - \frac{0.9028}{s} + \frac{0.8333}{s^2}$$

$$= \frac{0.0872s + 0.1566}{(s + 2.1197)^2 + (3.3589)^2} + \frac{0.8157}{s + 0.7607} - \frac{0.9028}{s} + \frac{0.8333}{s^2}$$

B-4-11. The equation of motion for the system is

$$m\ddot{x} + b_1\dot{x} + (k_1 + k_2)x = 0$$

By substituting the numerical values of m, k_1, k_2, and b_1 into this equation, we obtain

$$\ddot{x} + 4\dot{x} + 16x = 0$$

Laplace transforming this equation, we get

$$[s^2X(s) - sx(0) - \dot{x}(0)] + 4[sX(s) - x(0)] + 16X(s) = 0$$

or

$$(s^2 + 4s + 16)X(s) = sx(0) + \dot{x}(0) + 4x(0)$$

Solving this equation for X(s),

$$X(s) = \frac{sx(0) + \dot{x}(0) + 4x(0)}{s^2 + 4s + 16}$$

Since $x(0) = 0.05$ and $\dot{x}(0) = 1$, X(s) becomes

$$X(s) = \frac{0.05s + 1.2}{s^2 + 4s + 16}$$

$$= \frac{0.05(s + 2)}{(s + 2)^2 + (2\sqrt{3})^2} + \frac{0.3175 \times 2\sqrt{3}}{(s + 2)^2 + (2\sqrt{3})^2}$$

The inverse Laplace transform of X(s) gives

$$x(t) = 0.05\ e^{-2t} \cos 2\sqrt{3}\ t + 0.3175\ e^{-2t} \sin 2\sqrt{3}\ t$$

This equation gives the time response x(t).

The response curve x(t) versus t can be obtained easily by use of MATLAB. Noting that X(s) can be written as

$$X(s) = \frac{0.05s^2 + 1.2s}{s^2 + 4s + 16}\ \frac{1}{s}$$

we may define

num = [0.05   1.2   0]

den = [1   4   16]

and use a step command. The following MATLAB program will generate the response curve x(t) versus t as shown on the next page.

```
>> num = [0.05    1.2    0];
>> den = [1    4    16];
>> t = 0:0.01:4;
>> x = step(num,den,t);
>> plot(t,x)
>> grid
>> title('Response x(t)')
>> xlabel('t (sec)')
>> ylabel('x(t)')
```

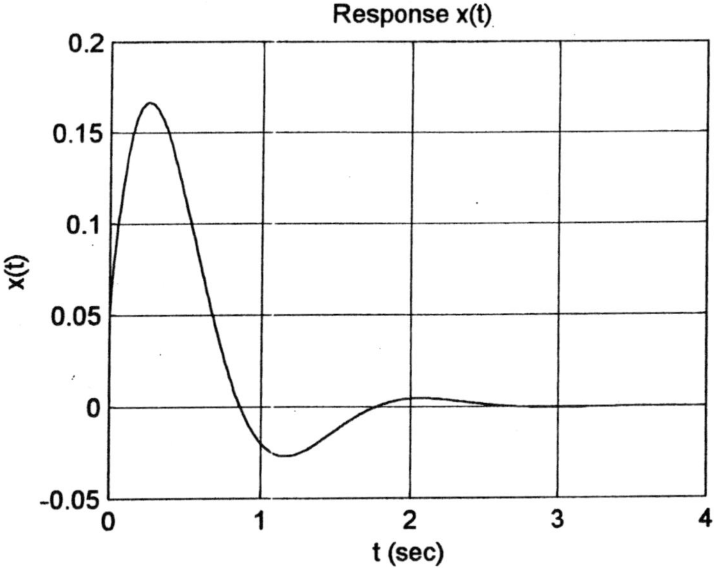

B-4-12. In this system u is the input and x_2 is the output. From Figure 6-57, we obtain the following equations:

$$b_1\dot{x}_1 + k_1(x_1 - x_2) = u$$

$$k_1(x_1 - x_2) = b_2\dot{x}_2 + k_2x_2$$

Laplace transforming these two equations, assuming zero initial conditions, we obtain

$$(b_1s + k_1)X_1(s) - k_1X_2(s) = U(s)$$

$$k_1X_1(s) = (b_2s + k_1 + k_2)X_2(s)$$

By eliminating $X_1(s)$ from these two equations, we get

$$(b_1s + k_1)\,\frac{b_2s + k_1 + k_2}{k_1}\,X_2(s) - k_1X_2(s) = U(s)$$

Simplifying this last equation, we get

$$[b_1b_2s^2 + (b_1k_1 + b_1k_2 + b_2k_1)s + k_1k_2]X_2(s) = k_1U(s)$$

from which we obtain

$$\frac{X_2(s)}{U(s)} = \frac{k_1}{b_1b_2s^2 + (b_1k_1 + b_1k_2 + b_2k_1)s + k_1k_2}$$

By substituting numerical values for k_1, k_2, b_1, and b_2 into this last equation, we obtain

$$\frac{X_2(s)}{U(s)} = \frac{4}{10s^2 + (4 + 20 + 40)s + 4 \times 20}$$

$$= \frac{0.4}{s^2 + 6.4s + 8}$$

Since the input u is a step force of 2N, we have U(s) = 2/s. $X_2(s)$ can be obtained from

$$X_2(s) = \frac{0.4}{s^2 + 6.4s + 8} \frac{2}{s}$$

$$= \frac{0.8}{(s + 4.6967)(s + 1.7033)s}$$

$$= \frac{0.1}{s} + \frac{0.0569}{s + 4.6967} + \frac{-0.1569}{s + 1.7033}$$

The inverse Laplace transform of $X_2(s)$ gives

$$x_2(t) = 0.1 + 0.0569\ e^{-4.6967t} - 0.1569\ e^{-1.7033t}$$

The response curve $x_2(t)$ versus t can be obtained with MATLAB as follows: First note that

$$X_2(s) = \frac{0.8}{s^2 + 6.4s + 8} \frac{1}{s}$$

Then, define

num = [0 0 0.8]

den = [1 6.4 8]

and use a step command. The following MATLAB program will yield the response curve $x_2(t)$. The resulting response curve is shown in the figure on the next page.

```
>> num = [0    0    0.8];
>> den = [1    6.4    8];
>> t = 0:0.01:4;
>> x = step(num,den,t);
>> plot(t,x)
>> grid
>> title('Response x(t)')
>> xlabel('t (sec)')
>> ylabel('x(t)')
```

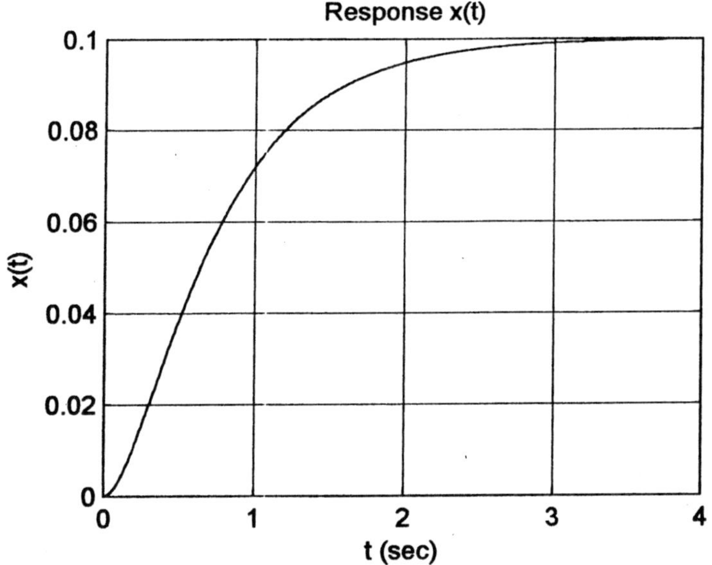

B-4-13. **The system equation is**

$$m\ddot{x} + b\dot{x} = \delta(t), \qquad x(0-) = 0, \qquad \dot{x}(0-) = 0$$

The $\mathcal{L}_-$ transform of this equation is

$$(ms^2 + bs)X(s) = 1$$

Solving for X(s), we get

$$X(s) = \frac{1}{ms^2 + bs} = \frac{1}{m} \frac{1}{s(s + \frac{b}{m})} = \frac{1}{b}\left(\frac{1}{s} - \frac{1}{s + \frac{b}{m}}\right)$$

By substituting the given numerical values for m and b, we obtain

$$X(s) = \frac{1}{200}\left(\frac{1}{s} - \frac{1}{s + 2}\right)$$

Then the response x(t) is obtained as

$$x(t) = 0.005(1 - e^{-2t})$$

The velocity $\dot{x}(t)$ is

$$\dot{x}(t) = 0.01\ e^{-2t}$$

Thus

$$\dot{x}(0+) = 0.01 \text{ m/s}$$

The initial velocity $\dot{x}(0+)$ can also be obtained by use of the initial-value theorem.

$$\dot{x}(0+) = \lim_{s \to \infty} s^2 X(s) = \lim_{s \to \infty} \frac{s^2}{ms^2 + bs} = \frac{1}{m} = 0.01$$

The response curve x(t) versus t is an exponential curve as shown below.

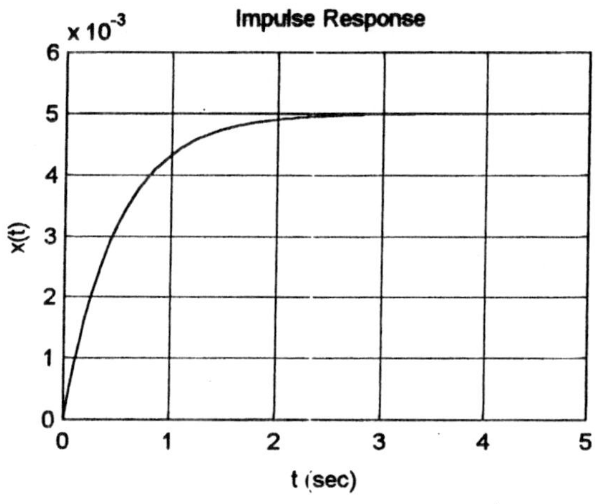

B-4-14. The system equation is

$$m\ddot{x} + kx = \delta(t)$$

Taking the $\mathcal{L}_-$ transform of the equation, using the initial conditions $x(0-) = 0$ and $\dot{x}(0-) = 0$, we obtain

$$(ms^2 + k)X(s) = \mathcal{L}_-[\delta(t)]$$

Hence, the transfer function $X(s)/\mathcal{L}_-[\delta(t)]$ of the system is

$$\frac{X(s)}{\mathcal{L}_-[\delta(t)]} = \frac{1}{ms^2 + k}$$

Since $\mathcal{L}_-[\delta(t)] = 1$, we have

$$X(s) = \frac{1}{ms^2 + k} = \frac{1}{\sqrt{km}} \frac{\sqrt{\frac{k}{m}}}{s^2 + \left(\sqrt{\frac{k}{m}}\right)^2}$$

The inverse Laplace transform of X(s) is

$$x(t) = \frac{1}{\sqrt{km}} \sin \sqrt{\frac{k}{m}}\, t$$

Thus, the response x(t) is a sinusoidal motion. The velocity $\dot{x}(t)$ is

$$\dot{x}(t) = \frac{1}{m} \cos \sqrt{\frac{k}{m}}\, t$$

Hence, the initial velocity $\dot{x}(0+)$ is

$$\dot{x}(0+) = \frac{1}{m}$$

B-4-15. Define the force input to the system as u(t). Then the system equation is

$$m\ddot{x} + b\dot{x} + kx = u(t)$$

where u(t) is a unit-impulse force $\delta(t)$. Taking the $\mathcal{L}$-transform of this equation with $x(0-) = 0$ and $\dot{x}(0-) = 0$, we obtain

$$(ms^2 + bs + k)X(s) = U(s)$$

The transfer function of the system is

$$\frac{X(s)}{U(s)} = \frac{1}{ms^2 + bs + k}$$

By substituting the given numerical values into the transfer function, we obtain

$$\frac{X(s)}{U(s)} = \frac{1}{10s^2 + 20s + 50}$$

The following MATLAB program produces the response x(t). The response curve is shown below.

```
>> num = [1];
>> den = [10    20    50];
>> t = 0:0.01:8;
>> impulse(num,den,t)
>> grid
>> xlabel('t')
>> ylabel('x(t)')
```

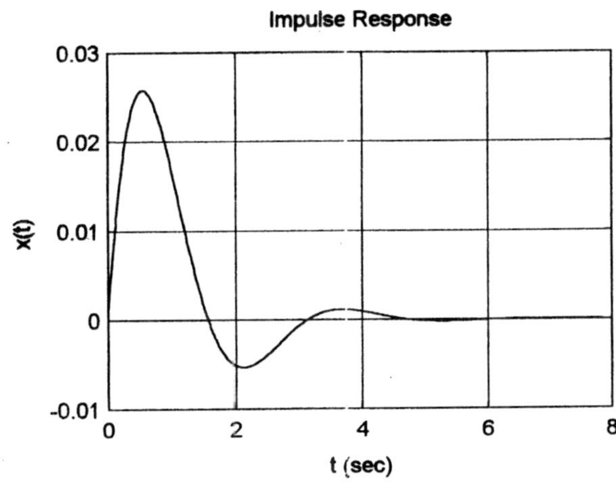

B-4-16. The equation of motion for the system is

$$m\ddot{x} + kx = p(t)$$

where m = 1 kg, k = 100 N/m, $p(t) = 10\,\delta(t)$ N, $x(0-) = 0.1$ m, and $\dot{x}(0-) = 1$ m/s. By substituting the given numerical values into the system equation, we obtain

$$\ddot{x} + 100\,x = 10\,\delta(t)$$

Taking the $\mathcal{L}_-$ transform of this last equation gives

$$[s^2X(s) - sx(0-) - \dot{x}(0-)] + 100\,X(s) = 10$$

or

$$(s^2 + 100)X(s) = 10 + 0.1s + 1 = 11 + 0.1s$$

Solving for X(s) gives

$$X(s) = \frac{11 + 0.1s}{s^2 + 10^2}$$

The inverse Laplace transform of X(s) gives

$$x(t) = \frac{11}{10}\sin 10t + 0.1\cos 10t$$

B-4-17. The system equation is

$$m\ddot{x} = -b_1\dot{x} - k_1x - b_2\dot{x} - k_2x$$

or

$$m\ddot{x} + (b_1 + b_2)\dot{x} + (k_1 + k_2)x = 0$$

Laplace transforming this last equation, we obtain

$$m[s^2X(s) - sx(0) - \dot{x}(0)] + (b_1 + b_2)[sX(s) - x(0)]$$
$$+ (k_1 + k_2)X(s) = 0$$

Rewriting, we have

$$[ms^2 + (b_1 + b_2)s + (k_1 + k_2)]X(s)$$
$$= msx(0) + m\dot{x}(0) + (b_1 + b_2)x(0)$$

Solving for X(s), we get

$$X(s) = \frac{mx(0)s + m\dot{x}(0) + (b_1 + b_2)x(0)}{ms^2 + (b_1 + b_2)s + (k_1 + k_2)}$$

By substituting $x(0) = x_o$, $\dot{x}(0) = v_o$ and the given numerical values into this last equation, we obtain

$$X(s) = \frac{10x_o s + 10v_o + 120x_o}{10s^2 + 120s + 1000}$$

$$= \frac{x_o s + v_o + 12x_o}{s^2 + 12s + 100}$$

$$= \frac{x_o(s + 6) + (v_o + 6x_o)}{(s + 6)^2 + 8^2}$$

The inverse Laplace transform of X(s) gives the output motion x(t).

$$x(t) = x_o e^{-6t} \cos 8t + \frac{v_o + 6x_o}{8} e^{-6t} \sin 8t$$

The motion x(t) is a damped sinusoidal motion.

B-4-18. Referring to the solution to Problem B-4-17, we have

$$X(s) = \frac{mx(0)s + m\dot{x}(0) + (b_1 + b_2)x(0)}{ms^2 + (b_1 + b_2)s + (k_1 + k_2)}$$

Substituting $x(0) = 0$ m, $\dot{x}(0) = 0.5$ m/s, $m = 100$ kg, $b_1 = 120$ N-s/m, $b_2 = 80$ N-s/m, $k_1 = 200$ N/m, and $k_2 = 300$ N/m into this last equation, we obtain

$$X(s) = \frac{100 \times 0.5}{100s^2 + 200s + 500}$$

$$= \frac{0.5}{s^2 + 2s + 5}$$

The response x(t) to the given initial condition can be obtained from X(s) by rewriting it as

$$X(s) = \frac{0.5s}{s^2 + 2s + 5} \frac{1}{s}$$

and using a step command. A possible MATLAB program to obtain the response x(t) is given below. The resulting response curve x(t) versus t is shown on the next page.

```
>> num = [0.5   0];
>> den = [1   2   5];
>> t = 0:0.01:6;
>> x = step(num,den,t);
>> plot(t,x)
>> grid
>> title('Response to Initial Condition');
>> xlabel('t (sec)')
>> ylabel('x(t)')
```

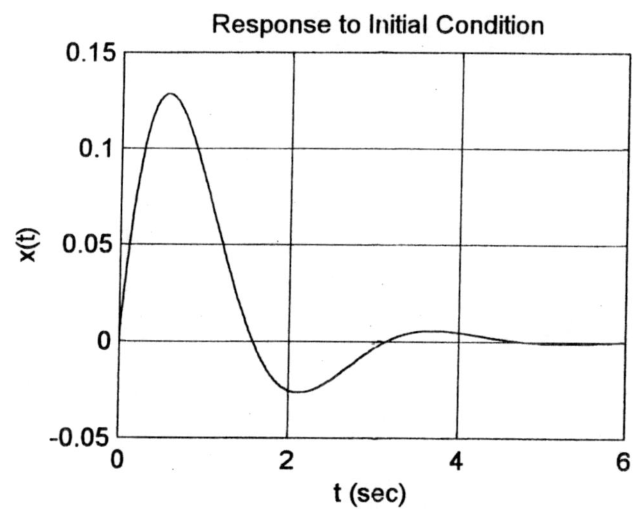
Response to Initial Condition
0.15
0.1
0.05
0
-0.05
x(t)
0
2
4
6
t (sec)

CHAPTER 5

B-5-1.

(a) For the system shown in Figure 5-36(a), the system equation is

$$m\ddot{x} + kx = u$$

Define state variables as

$$x_1 = x$$

$$x_2 = \dot{x}$$

and the output variable y as

$$y = x$$

Then a state space representation of the system is

$$\dot{x}_1 = x_2$$

$$\dot{x}_2 = -\frac{k}{m} x_1 + \frac{1}{m} u$$

The output equation is

$$y = x_1$$

In terms ofthe vector-matrix expression,

$$\begin{bmatrix} \dot{x}_1 \\ \dot{x}_2 \end{bmatrix} = \begin{bmatrix} 0 & 1 \\ -\frac{k}{m} & 0 \end{bmatrix} \begin{bmatrix} x_1 \\ x_2 \end{bmatrix} + \begin{bmatrix} 0 \\ \frac{1}{m} \end{bmatrix} u$$

$$y = [1 \quad 0] \begin{bmatrix} x_1 \\ x_2 \end{bmatrix}$$

(b) For the system shown in Figure 5-36(b), define the displacement of a point between two springs as z. Then we obtain the following equations for the system:

$$m\ddot{x} = k_2(z - x) + u \qquad (1)$$

$$k_2(x - z) = k_1 z \qquad (2)$$

From Equation (2) we have

$$z = \frac{k_2}{k_1 + k_2} x \tag{3}$$

By substituting Equation (3) into Equation (1), we obtain

$$m\ddot{x} = -\frac{k_1 k_2}{k_1 + k_2} x + u \tag{4}$$

Now define state variables x_1 and x_2 as

$$x_1 = x$$

$$x_2 = \dot{x}$$

and the output variable y as

$$y = x$$

Then, Equation (4) can be written as

$$m\dot{x}_2 = -\frac{k_1 k_2}{k_1 + k_2} x_1 + u$$

Then a state-space representation for the system is

$$\begin{bmatrix} \dot{x}_1 \\ \dot{x}_2 \end{bmatrix} = \begin{bmatrix} 0 & 1 \\ -\dfrac{k_1 k_2}{m(k_1 + k_2)} & 0 \end{bmatrix} \begin{bmatrix} x_1 \\ x_2 \end{bmatrix} + \begin{bmatrix} 0 \\ \dfrac{1}{m} \end{bmatrix} u$$

$$y = [1 \quad 0] \begin{bmatrix} x_1 \\ x_2 \end{bmatrix}$$

B-5-2. Define the tension acting in the wire as T. The equations of motion for the system are

$$J\ddot{\theta} = TR - k\theta R^2 \tag{1}$$

$$m\ddot{y} = -T + u \tag{2}$$

Simce $y = R\theta$, Equation (1) can be written as

$$J\ddot{y} = TR^2 - kR^2 y$$

By eliminating T from Equations (1) and (2), we have

$$J\ddot{y} = (u - m\ddot{y})R^2 - kR^2 y$$

or

$$(J + mR^2)\ddot{y} + kR^2 y = R^2 u$$

Thus,

$$\ddot{y} + \frac{kR^2}{J + mR^2} y = \frac{R^2}{J + mR^2} u$$

Define state variables x_1 and x_2 as

$$x_1 = y$$

$$x_2 = \dot{y}$$

Then the output y can be given by

$$y = x_1$$

Hence a state-space representation of the system is given by

$$\begin{bmatrix} \dot{x}_1 \\ \dot{x}_2 \end{bmatrix} = \begin{bmatrix} 0 & 1 \\ -\dfrac{kR^2}{J + mR^2} & 0 \end{bmatrix} \begin{bmatrix} x_1 \\ x_2 \end{bmatrix} + \begin{bmatrix} 0 \\ \dfrac{R^2}{J + mR^2} \end{bmatrix} u$$

$$y = [1 \quad 0] \begin{bmatrix} x_1 \\ x_2 \end{bmatrix}$$

B-5-3. The equations of motion for the system are

$$m_1\ddot{z} = -b_1(\dot{z} - \dot{y}) - k_1(z - y) + u$$

$$m_2\ddot{y} = -b_1(\dot{y} - \dot{z}) - k_1(y - z) - k_2 y$$

Rewriting these two equations, we obtain

$$m_1\ddot{z} + b_1\dot{z} + k_1 z = b_1\dot{y} + k_1 y + u \qquad (1)$$

$$m_2\ddot{y} + b_1\dot{y} + k_1 y + k_2 y = b_1\dot{z} + k_1 z \qquad (2)$$

Define state variables as follows:

$$x_1 = z$$

$$x_2 = \dot{z}$$

$$x_3 = y$$

$$x_4 = \dot{y}$$

Then, from Equation (1) we obtain

$$m_1\dot{x}_2 = -k_1x_1 - b_1x_2 + k_1x_3 + b_1x_4 + u$$

Also, from Equation (2) we get

$$m_2\dot{x}_4 = k_1x_1 + b_1x_2 - (k_1 + k_2)x_3 - b_1x_4$$

Hence, we obtain a state equation as follows:

$$\begin{bmatrix} \dot{x}_1 \\ \dot{x}_2 \\ \dot{x}_3 \\ \dot{x}_4 \end{bmatrix} = \begin{bmatrix} 0 & 1 & 0 & 0 \\ -\dfrac{k_1}{m_1} & -\dfrac{b_1}{m_1} & \dfrac{k_1}{m_1} & \dfrac{b_1}{m_1} \\ 0 & 0 & 0 & 1 \\ \dfrac{k_1}{m_2} & \dfrac{b_1}{m_2} & -\dfrac{k_1 + k_2}{m_2} & -\dfrac{b_1}{m_2} \end{bmatrix} \begin{bmatrix} x_1 \\ x_2 \\ x_3 \\ x_4 \end{bmatrix} + \begin{bmatrix} 0 \\ \dfrac{1}{m_1} \\ 0 \\ 0 \end{bmatrix} u \qquad (3)$$

The outputs of the system are z and y. Hence, if we define the output variables as

$$y_1 = z$$

$$y_2 = y$$

then

$$y_1 = x_1$$

$$y_2 = x_3$$

The output equation can now be put in the form

$$\begin{bmatrix} y_1 \\ y_2 \end{bmatrix} = \begin{bmatrix} 1 & 0 & 0 & 0 \\ 0 & 0 & 1 & 0 \end{bmatrix} \begin{bmatrix} x_1 \\ x_2 \\ x_3 \\ x_4 \end{bmatrix} \qquad (4)$$

Equations (3) and (4) give a state-space representation of the given system.

B-5-4. The equations of motion for the system are

$$m_1\ddot{y}_1 = -k_1y_1 - k_2(y_1 - y_2) + u_1$$

$$m_2\ddot{y}_2 = k_2(y_1 - y_2) - k_3y_2 + u_2$$

Rewriting these equations,

$$m_1\ddot{y}_1 = -(k_1 + k_2)y_1 + k_2y_2 + u_1 \qquad (1)$$

$$m_2\ddot{y}_2 = k_2y_1 - (k_2 + k_3)y_2 + u_2 \qquad (2)$$

Define state variables as follows:

$$x_1 = y_1$$

$$x_2 = \dot{y}_1$$

$$x_3 = y_2$$

$$x_4 = \dot{y}_2$$

Then, Equation (1) can be written as

$$m_1\dot{x}_2 = -(k_1 + k_2)x_1 + k_2x_3 + u_1$$

and Equation (2) becomes

$$m_2\dot{x}_4 = k_2x_1 - (k_2 + k_3)x_3 + u_2$$

Therefore, the state equation can be obtained as

$$\begin{bmatrix} \dot{x}_1 \\ \dot{x}_2 \\ \dot{x}_3 \\ \dot{x}_4 \end{bmatrix} = \begin{bmatrix} 0 & 1 & 0 & 0 \\ -\dfrac{k_1 + k_2}{m_1} & 0 & \dfrac{k_2}{m_1} & 0 \\ 0 & 0 & 0 & 1 \\ \dfrac{k_2}{m_2} & 0 & -\dfrac{k_2 + k_3}{m_2} & 0 \end{bmatrix} \begin{bmatrix} x_1 \\ x_2 \\ x_3 \\ x_4 \end{bmatrix} + \begin{bmatrix} 0 & 0 \\ \dfrac{1}{m_1} & 0 \\ 0 & 0 \\ 0 & \dfrac{1}{m_2} \end{bmatrix} \begin{bmatrix} u_1 \\ u_2 \end{bmatrix}$$

The output equation can be obtained as

$$\begin{bmatrix} y_1 \\ y_2 \end{bmatrix} = \begin{bmatrix} x_1 \\ x_3 \end{bmatrix} = \begin{bmatrix} 1 & 0 & 0 & 0 \\ 0 & 0 & 1 & 0 \end{bmatrix} \begin{bmatrix} x_1 \\ x_2 \\ x_3 \\ x_4 \end{bmatrix}$$

B-5-5. We shall first obtain the transfer function of the system from the given state-space equations. Referring to Equation (5-9):

$$G(s) = \mathbf{C}(s\mathbf{I} - \mathbf{A})^{-1}\mathbf{B} + D$$

we obtain

$$G(s) = [1 \quad 0 \quad 0] \begin{bmatrix} s & -1 & 0 \\ 0 & s & -1 \\ -1 & 3 & s - 3 \end{bmatrix}^{-1} \begin{bmatrix} 1 \\ 1 \\ 1 \end{bmatrix} + 0$$

Simplifying, we get

$$G(s) = \frac{1}{s^3 - 3s^2 + 3s - 1} [1 \quad 0 \quad 0] \begin{bmatrix} s^2 - 3s + 3 & s - 3 & 1 \\ 1 & s^2 - 3s & s \\ s & -3s + 1 & s^2 \end{bmatrix} \begin{bmatrix} 1 \\ 1 \\ 1 \end{bmatrix}$$

$$= \frac{1}{s^3 - 3s^2 + 3s - 1} (s^2 - 3s + 3 + s - 3 + 1)$$

$$= \frac{s^2 - 2s + 1}{s^3 - 3s^2 + 3s - 1}$$

Hence

$$\frac{Y(s)}{U(s)} = \frac{s^2 - 2s + 1}{s^3 - 3s^2 + 3s - 1}$$

from which we get

$$\dddot{y} - 3\ddot{y} + 3\dot{y} - y = \ddot{u} - 2\dot{u} + u$$

B-5-6. For the given system

$$\dddot{y} + 6\ddot{y} + 11\dot{y} + 6y = 6u$$

define state variables as

$$x_1 = y$$
$$x_2 = \dot{y}$$
$$x_3 = \ddot{y}$$

Then

$$\dot{x}_1 = x_2$$
$$\dot{x}_2 = x_3$$
$$\dot{x}_3 = \dddot{y} = -6y - 11\dot{y} - 6\ddot{y} + 6u$$
$$= -6x_1 - 11x_2 - 6x_3 + 6u$$

Hence, we obtain

$$\begin{bmatrix} \dot{x}_1 \\ \dot{x}_2 \\ \dot{x}_3 \end{bmatrix} = \begin{bmatrix} 0 & 1 & 0 \\ 0 & 0 & 1 \\ -6 & -11 & -6 \end{bmatrix} \begin{bmatrix} x_1 \\ x_2 \\ x_3 \end{bmatrix} + 6u$$

$$y = [1 \quad 0 \quad 0] \begin{bmatrix} x_1 \\ x_2 \\ x_3 \end{bmatrix}$$

B-5-7. Referring to Equation (5-9), we have

$$G(s) = \underset{\sim}{C}(s\underset{\sim}{I} - \underset{\sim}{A})^{-1}\underset{\sim}{B} + D$$

Hence

$$G(s) = [1 \quad 0] \begin{bmatrix} s + 4 & 1 \\ -3 & s + 1 \end{bmatrix}^{-1} \begin{bmatrix} 1 \\ 1 \end{bmatrix} + 0$$

$$= [1 \quad 0] \frac{1}{(s + 4)(s + 1) + 3} \begin{bmatrix} s + 1 & -1 \\ 3 & s + 4 \end{bmatrix} \begin{bmatrix} 1 \\ 1 \end{bmatrix}$$

$$= \frac{s}{s^2 + 5s + 7}$$

B-5-8. Referring to Equation (5-9), we have

$$G(s) = \underset{\sim}{C}(s\underset{\sim}{I} - \underset{\sim}{A})^{-1}\underset{\sim}{B} + D$$

Hence

$$G(s) = [1 \quad 0] \begin{bmatrix} s & -1 \\ 1 & s + 2 \end{bmatrix}^{-1} \begin{bmatrix} 1 \\ 1 \end{bmatrix} + 1$$

$$= [1 \quad 0] \frac{1}{s^2 + 2s + 1} \begin{bmatrix} s + 2 & 1 \\ -1 & s \end{bmatrix} \begin{bmatrix} 1 \\ 1 \end{bmatrix} + 1$$

$$= \frac{s + 3}{s^2 + 2s + 1} + 1$$

$$= \frac{s^2 + 3s + 4}{s^2 + 2s + 1}$$

B-5-9. We use MATLAB to obtain the transfer function of the system. The following MATLAB program produces the transfer function:

```
>> A = [0    1    0;0    0    1;-600    -100    -10];
>> B = [0;10;0];
>> C = [1    0    0];
>> D = 0;
>> [num,den] = ss2tf(A,B,C,D)

num =

         0    0.0000   10.0000  100.0000


den =

    1.0000   10.0000  100.0000  600.0000
```

From the MATLAB output, we get

$$G(s) = \frac{10s + 100}{s^3 + 10s^2 + 100s + 600}$$

B-5-10. The MATLAB program given below produces the unit-step response curves. The resulting response curves are shown below.

```
>> A = [-1    -1;5    0];
>> B = [1    1;1    0];
>> C = [1    0;0    1];
>> D = [1    0;0    0];
>> sys = ss(A,B,C,D);
>> step(sys)
>> grid
>> title('Unit-Step Responses')
>> xlabel('t')
>> ylabel('Outputs')
```

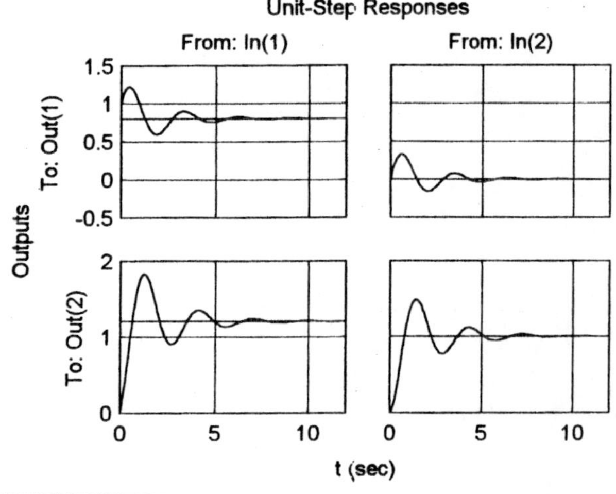

B-5-11. The following MATLAB program produces the unit-step response curve and unit-impulse response curve. The resulting two curves are shown below.

```
>> A = [-5   -25   -5;1   0   0;0   1   0];
>> B = [1;0;0];
>> C = [0   25   5];
>> D = 0;
>> sys = ss(A,B,C,D);
>> subplot(221); step(sys)
>> grid
>> title('Unit-Step Response')
>> xlabel('t')
>> ylabel('y(t)')
>> subplot(222); impulse(sys)
>> grid
>> title('Unit-Impulse Response')
>> xlabel('t')
>> ylabel('y(t)')
```

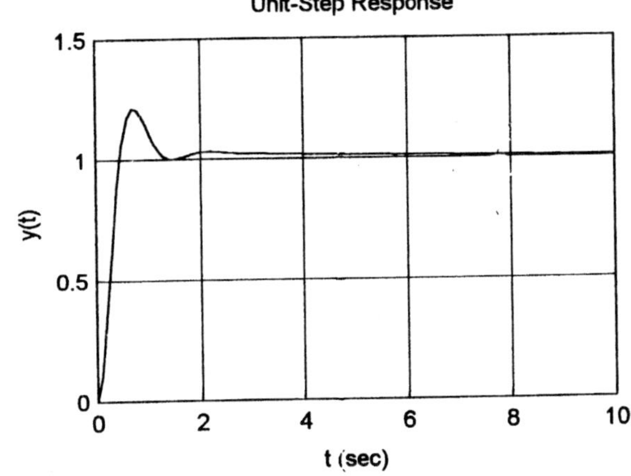

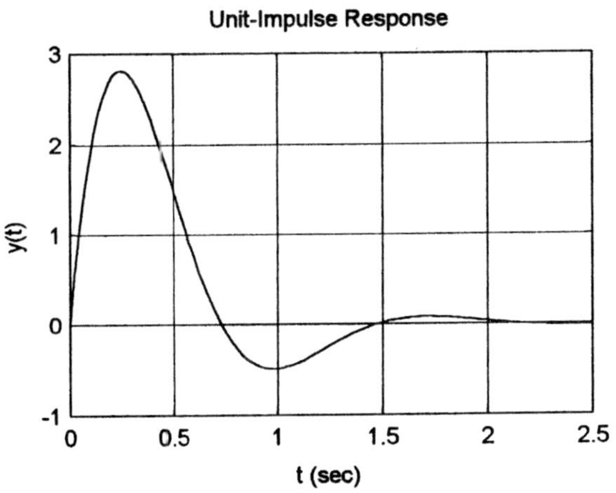

B-5-12. The following MATLAB program produces the response of the system to the given initial condition. The response curve is shown on the next page.

```
>> t = 0:0.01:6;
>> A = [0   1;-10   -5];
>> B = [0;0];
>> C = [1  0];
>> D = 0;
>> [y] = initial(A,B,C,D,[2;1],t);
>> plot(t,y)
>> grid
>> title('Response to Initial Condition')
>> xlabel('t (sec)')
>> ylabel('y(t)')
```

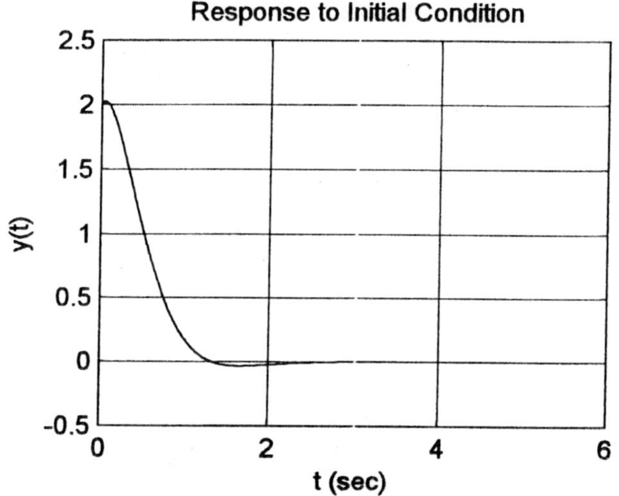

B-5-13. Let us assume that the input to the system is u. (u = 0 in this problem.) Then the system equation becomes

$$\dddot{y} + 8\ddot{y} + 17\dot{y} + 10y = u$$

Define state variables as

$$x_1 = y$$
$$x_2 = \dot{y}$$
$$x_3 = \ddot{y}$$

Then we obtain

$$\dot{x}_1 = x_2$$
$$\dot{x}_2 = x_3$$
$$\dot{x}_3 = -10x_1 - 17x_2 - 8x_3 + u$$

or

$$\begin{bmatrix} \dot{x}_1 \\ \dot{x}_2 \\ \dot{x}_3 \end{bmatrix} = \begin{bmatrix} 0 & 1 & 0 \\ 0 & 0 & 1 \\ -10 & -17 & -8 \end{bmatrix} \begin{bmatrix} x_1 \\ x_2 \\ x_3 \end{bmatrix} + \begin{bmatrix} 0 \\ 0 \\ 1 \end{bmatrix} u$$

$$y = [1 \quad 0 \quad 0] \begin{bmatrix} x_1 \\ x_2 \\ x_3 \end{bmatrix}$$

The following MATLAB program produces the response to the given initial condition. The response curve is shown below.

```
>> t = 0:0.01:10;
>> A = [0  1  0;0  0  1;-10  -17  -8];
>> B = [0;0;0];
>> C = [1  0  0];
>> D = 0;
>> sys = ss(A,B,C,D);
>> u = t;
>> [y,t] = lsim(sys,u,t,[2;1;0.5]);
>> plot(t,y)
>> grid
>> title('Response to Initial Condition')
>> xlabel('t (sec)')
>> ylabel('y(t)')
```

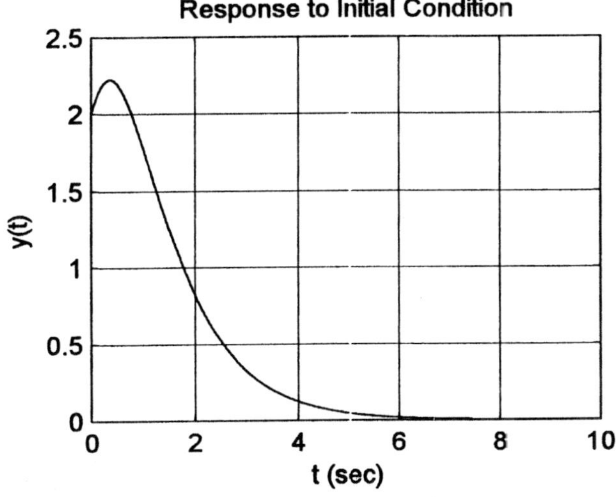

B-5-14. The equation of motion for the system is

$$m\ddot{y} + b\dot{y} + ky = u$$

By substituting the given numerical values for m, b, and k, we obtain

$$5\ddot{y} + 8\dot{y} + 20y = u$$

Define state variables as follows

$$x_1 = y$$

$$x_2 = \dot{y}$$

Then,

$$\dot{x}_1 = x_2$$

$$\dot{x}_2 = -4x_1 - 1.6x_2 + 0.2u$$

The state-space representation for the system is

$$\begin{bmatrix} \dot{x}_1 \\ \dot{x}_2 \end{bmatrix} = \begin{bmatrix} 0 & 1 \\ -4 & -1.6 \end{bmatrix} \begin{bmatrix} x_1 \\ x_2 \end{bmatrix} + \begin{bmatrix} 0 \\ 0.2 \end{bmatrix} u$$

$$y = [1 \quad 0] \begin{bmatrix} x_1 \\ x_2 \end{bmatrix}$$

The MATLAB program presented on the next page produces the response curve y(t) versus t and the input curve u(t) versus t.

```
>> t = 0:0.01:15;
>> A = [0   1;-4   -1.6];
>> B = [0;0.2];
>> C = [1   0];
>> D = 0;
>> sys = ss(A,B,C,D);
>> t1 = 0:0.01:5;
>> u1 = [1*ones(size(t1))];
>> u2 = 0*[5.01:0.01:15];
>> u = [u1   u2];
>> y = lsim(sys,u,t);
>> subplot(221); plot(t,y)
>> grid
>> title('Response to u(t)')
>> xlabel('t (sec)')
>> ylabel('y(t)')
>> subplot(222); plot(t,u,t,u,'.')
>> v = [-1   15   -1   2]; axis(v)
>> grid
>> title('Input u(t)')
>> xlabel('t (sec)')
>> ylabel('u(t)')
```

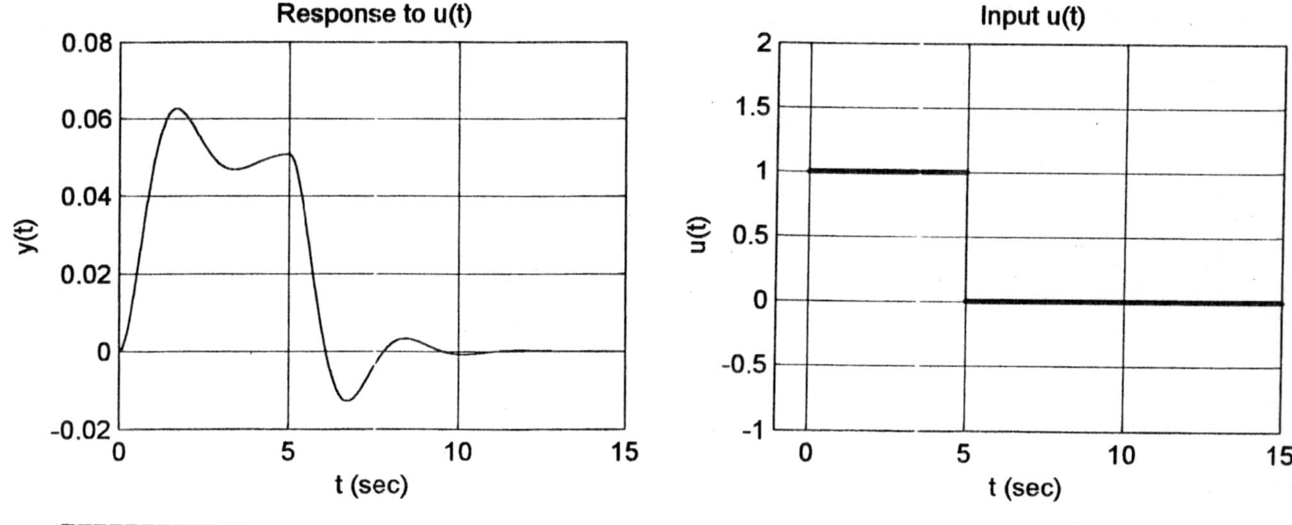

B-5-15. The system equation for $t > 0$ is

$$m\ddot{y} + b\dot{y} + ky = mg$$

By substituting the given numerical values for m, b, and k into this equation, we get

$$\ddot{y} + 4\dot{y} + 40y = 9.807$$

To derive a state-space equation for $t > 0$, define the input force to the system as u. Then the system equation becomes

$$\ddot{y} + 4\dot{y} + 40y = u$$

Define state variables as follows:

$$x_1 = y$$

$$x_2 = \dot{y}$$

Then the state-space representation of the system can be given as

$$\begin{bmatrix} \dot{x}_1 \\ \dot{x}_2 \end{bmatrix} = \begin{bmatrix} 0 & 1 \\ -40 & -4 \end{bmatrix} \begin{bmatrix} x_1 \\ x_2 \end{bmatrix} + \begin{bmatrix} 0 \\ 1 \end{bmatrix} u$$

$$y = \begin{bmatrix} 1 & 0 \end{bmatrix} \begin{bmatrix} x_1 \\ x_2 \end{bmatrix}$$

In this problem the force mg acts as the input force. Thus,

$$u = mg = 9.807 \text{ kg-m/s}^2 = 9.807 \text{ N}$$

The state-space representation becomes

$$\begin{bmatrix} \dot{x}_1 \\ \dot{x}_2 \end{bmatrix} = \begin{bmatrix} 0 & 1 \\ -40 & -4 \end{bmatrix} \begin{bmatrix} x_1 \\ x_2 \end{bmatrix} + \begin{bmatrix} 0 \\ 9.807 \end{bmatrix} 1(t)$$

$$y = \begin{bmatrix} 1 & 0 \end{bmatrix} \begin{bmatrix} x_1 \\ x_2 \end{bmatrix}$$

The MATLAB program below produces the response curve y(t) versus t.

```
>> A = [0   1;-40   -4];
>> B = [0;9.807];
>> C = [1   0];
>> D = 0;
>> sys = ss(A,B,C,D);
>> step(sys)
>> grid
>> title('Response to Mass m placed on massless bar')
>> xlabel('t')
>> ylabel('y(t)')
```

The response curve is shown below.

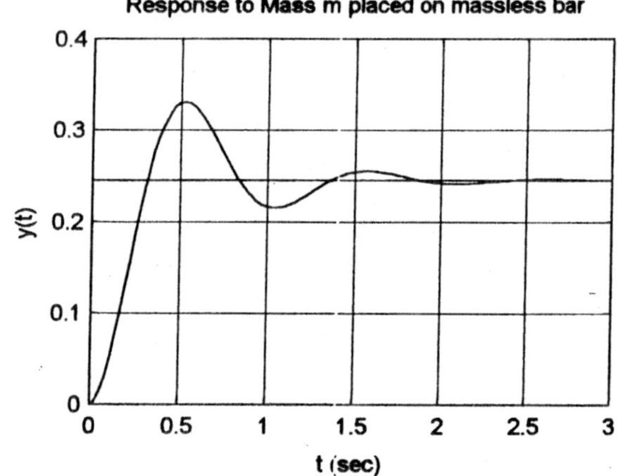

B-5-16. The system equation is

$$m\ddot{y} = -ky - b(\dot{y} - \dot{u})$$

or

$$m\ddot{y} + b\dot{y} + ky = b\dot{u}$$

which can be rewritten as

$$\ddot{y} + \frac{b}{m}\dot{y} + \frac{k}{m}y = \frac{b}{m}\dot{u} \qquad (1)$$

To obtain a state-space representation of this system, we shall use Method 1 presented in Section 5-4. Let us compare Equation (1) with the standard form

$$\ddot{y} + a_1\dot{y} + a_2 y = b_0\ddot{u} + b_1\dot{u} + b_2 u$$

and identify

$$a_1 = \frac{b}{m}, \quad a_2 = \frac{k}{m}, \quad b_0 = 0, \quad b_1 = \frac{b}{m}, \quad b_2 = 0$$

From Equations (5-23), (5-24) and (5-29) we have

$$\beta_0 = b_0 = 0$$

$$\beta_1 = b_1 - a_1\beta_0 = \frac{b}{m}$$

$$\beta_2 = b_2 - a_1\beta_1 - a_2\beta_0 = -\left(\frac{b}{m}\right)^2$$

Referring to Equations (5-21) and (5-22), we define

$$x_1 = y - \beta_0 u = y$$

$$x_2 = \dot{x}_1 - \beta_1 u = \dot{x}_1 - \frac{b}{m} u \tag{2}$$

From Equation (2) we obtain

$$\dot{x}_1 = x_2 + \frac{b}{m} u$$

From Equation (5-28) we have

$$\dot{x}_2 = - a_2 x_1 - a_1 x_2 + \beta_2 u$$

or

$$\dot{x}_2 = - \frac{k}{m} x_1 - \frac{b}{m} x_2 - \left(\frac{b}{m}\right)^2 u$$

Hence, a state-space representation of the present system is

$$\begin{bmatrix} \dot{x}_1 \\ \dot{x}_2 \end{bmatrix} = \begin{bmatrix} 0 & 1 \\ -\frac{k}{m} & -\frac{b}{m} \end{bmatrix} \begin{bmatrix} x_1 \\ x_2 \end{bmatrix} + \begin{bmatrix} \frac{b}{m} \\ -\left(\frac{b}{m}\right)^2 \end{bmatrix} u$$

$$y = [1 \quad 0] \begin{bmatrix} x_1 \\ x_2 \end{bmatrix}$$

By substituting the given numerical values for m, b, and k, the state equation becomes

$$\begin{bmatrix} \dot{x}_1 \\ \dot{x}_2 \end{bmatrix} = \begin{bmatrix} 0 & 1 \\ -4 & -2 \end{bmatrix} \begin{bmatrix} x_1 \\ x_2 \end{bmatrix} + \begin{bmatrix} 2 \\ -4 \end{bmatrix} u$$

Since $u = 0.2 \cdot 1(t)$, the state equation can be rewritten as

$$\begin{bmatrix} \dot{x}_1 \\ \dot{x}_2 \end{bmatrix} = \begin{bmatrix} 0 & 1 \\ -4 & -2 \end{bmatrix} \begin{bmatrix} x_1 \\ x_2 \end{bmatrix} + \begin{bmatrix} 0.4 \\ -0.8 \end{bmatrix} 1(t)$$

The output equation is

$$y = [1 \quad 0] \begin{bmatrix} x_1 \\ x_2 \end{bmatrix}$$

The MATLAB program presented on the next page produces the response y(t). The response curve is shown also on the next page.

```
>> t = 0:0.01:6;
>> A = [0   1;-4   -2];
>> B = [0.4;-0.8];
>> C = [1   0];
>> D = 0;
>> sys = ss(A,B,C,D);
>> step(sys,t)
>> grid
>> xlabel('t')
>> ylabel('y(t)')
```

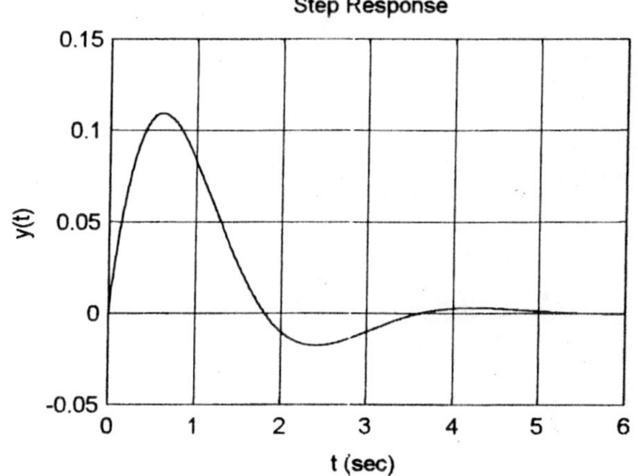

Notice that although this is a step response, the response start from $y(0) = 0$ and end at zero, or $y(\infty) = 0$.

B-5-17. The system equations are

$$m_1\ddot{z}_1 = -b(\dot{z}_1 - \dot{z}_2) - k(z_1 - z_2)$$

$$m_2\ddot{z}_2 = -b(\dot{z}_2 - \dot{z}_1) - k(z_2 - z_1) + f$$

Rewriting,

$$m_1\ddot{z}_1 + b\dot{z}_1 + kz_1 = b\dot{z}_2 + kz_2 \qquad (1)$$

$$m_2\ddot{z}_2 + b\dot{z}_2 + kz_2 = b\dot{z}_1 + kz_1 + f \qquad (2)$$

Define

$$z_2 - z_1 = z$$

and define state variables as

$$x_1 = z$$

$$x_2 = \dot{z}$$

$$x_3 = z_1$$

$$x_4 = \dot{z}_1$$

By eliminating z_2 from Equations (1) and (2), we obtain

$$m_1\ddot{z}_1 = b\dot{z} + kz \qquad (3)$$

$$m_2(\ddot{z}_1 + \ddot{z}) + b\dot{z} + kz = f \qquad (4)$$

In terms of state variables, Equations (3) and (4) can be written as

$$m_1\dot{x}_4 = bx_2 + kx_1 \qquad (5)$$

$$m_2\dot{x}_4 + m_2\dot{x}_2 + bx_2 + kx_1 = f \qquad (6)$$

Eliminating $\dot{x}_4$ from Equations (5) and (6), we obtain

$$\dot{x}_2 = -\frac{(m_1 + m_2)b}{m_1m_2}x_2 - \frac{(m_1 + m_2)k}{m_1m_2}x_1 + \frac{1}{m_2}f$$

Since

$$\dot{x}_1 = x_2$$

$$\dot{x}_3 = x_4$$

and from Equation (5),

$$\dot{x}_4 = \frac{k}{m_1}x_1 + \frac{b}{m_1}x_2$$

we obtain the following state equation:

$$\begin{bmatrix} \dot{x}_1 \\ \dot{x}_2 \\ \dot{x}_3 \\ \dot{x}_4 \end{bmatrix} = \begin{bmatrix} 0 & 1 & 0 & 0 \\ -\frac{(m_1 + m_2)k}{m_1m_2} & -\frac{(m_1 + m_2)b}{m_1m_2} & 0 & 0 \\ 0 & 0 & 0 & 1 \\ \frac{k}{m} & \frac{b}{m} & 0 & 0 \end{bmatrix} \begin{bmatrix} x_1 \\ x_2 \\ x_3 \\ x_4 \end{bmatrix} + \begin{bmatrix} 0 \\ \frac{1}{m_2} \\ 0 \\ 0 \end{bmatrix} f$$

Considering $z = x_1$ and $z_1 = x_3$ as the outputs, we have

$$\begin{bmatrix} y_1 \\ y_2 \end{bmatrix} = \begin{bmatrix} 1 & 0 & 0 & 0 \\ 0 & 0 & 1 & 0 \end{bmatrix} \begin{bmatrix} x_1 \\ x_2 \\ x_3 \\ x_4 \end{bmatrix} + \begin{bmatrix} 0 \\ 0 \end{bmatrix} f$$

Define a force input of 1 N as u. Then f = 10 N = 10u. By substituting the numerical values for m_1, m_2, b, and k into the state equation and f = 10u into both the state equation and output equation, we obtain

$$\begin{bmatrix} \dot{x}_1 \\ \dot{x}_2 \\ \dot{x}_3 \\ \dot{x}_4 \end{bmatrix} = \begin{bmatrix} 0 & 1 & 0 & 0 \\ -9 & -3 & 0 & 0 \\ 0 & 0 & 0 & 1 \\ 6 & 2 & 0 & 0 \end{bmatrix} \begin{bmatrix} x_1 \\ x_2 \\ x_3 \\ x_4 \end{bmatrix} + \begin{bmatrix} 0 \\ 0.5 \\ 0 \\ 0 \end{bmatrix} u$$

$$\begin{bmatrix} y_1 \\ y_2 \end{bmatrix} = \begin{bmatrix} 1 & 0 & 0 & 0 \\ 0 & 0 & 1 & 0 \end{bmatrix} \begin{bmatrix} x_1 \\ x_2 \\ x_3 \\ x_4 \end{bmatrix} + \begin{bmatrix} 0 \\ 0 \end{bmatrix} u$$

The following MATLAB program produces the required responses $z_1(t)$, $z_2(t)$, and $z(t)$. [Note that $z_1(t) = x_3(t)$, $z_2(t) = z_1(t) + z(t) = x_3(t) + x_1(t)$ and $z(t) = x_1(t)$.]
The response curves $z_1(t)$ versus t and $z_2(t)$ versus t are shown below. The response curve z(t) versus t is shown on the next page.

```
>> t = 0:0.01:15;
>> A = [0  1  0  0;-9  -3  0  0;0  0  0  1;6  2  0  0];
>> B = [0;0.5;0;0];
>> C = [1  0  0 0;0  0  1  0];
>> D = [0;0];
>> sys = ss(A,B,C,D);
>> [y,t,x] = step(sys,t);
>> x1 = [1  0  0  0]*x';
>> x2 = [0  1  0  0]*x';
>> x3 = [0  0  1  0]*x';
>> x4 = [0  0  0  1]*x';
>> subplot(221); plot(t,x3); grid
>> xlabel('t (sec)'); ylabel('x3 = z_1')
>> subplot(222); plot(t,x1+x3); grid
>> xlabel('t (sec)'); ylabel('x1 + x3 = z_2')
>> subplot(223); plot(t,x1); grid
>> xlabel('t (sec)'); ylabel('x1 = z')
```

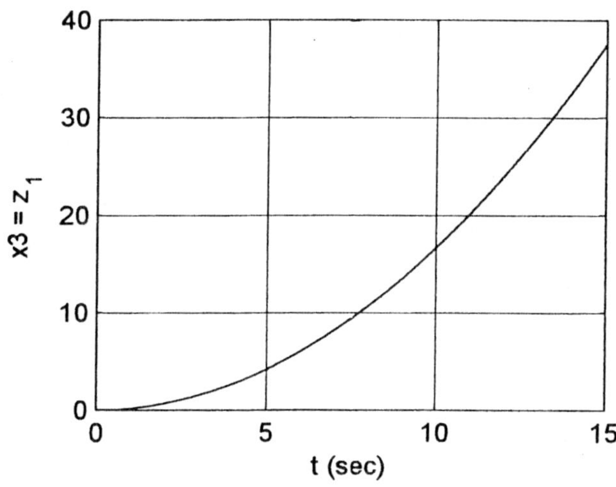

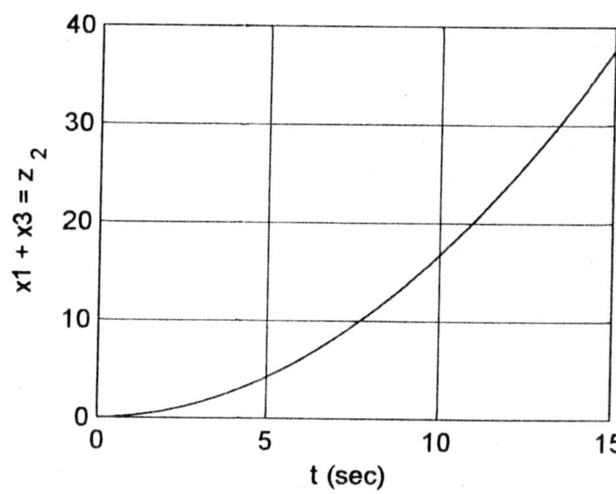

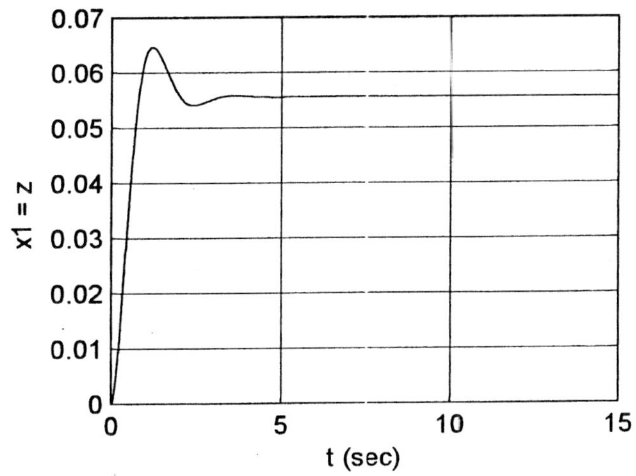

B-5-18. The system equations are

$$m_1\ddot{z}_1 = -k_1 z_1 - b(\dot{z}_1 - \dot{z}_2) - k_2(z_1 - z_2) \qquad (1)$$

$$m_2\ddot{z}_2 = -b(\dot{z}_2 - \dot{z}_1) - k_2(z_2 - z_1) + f \qquad (2)$$

Define state variables as follows:

$$x_1 = z_1$$

$$x_2 = \dot{z}_1$$

$$x_3 = z_2$$

$$x_4 = \dot{z}_2$$

Then, Equations (1) and (2) can be written as

$$m_1\dot{x}_2 = -k_1x_1 - bx_2 + bx_4 - k_2x_1 + k_2x_3$$

$$m_2\dot{x}_4 = -bx_4 + bx_2 - k_2x_3 + k_2x_1 + f$$

Hence the state equation can be obtained as follows:

$$\dot{x}_1 = x_2$$

$$\dot{x}_2 = -\frac{(k_1 + k_2)}{m_1} x_1 - \frac{b}{m_1} x_2 + \frac{k_2}{m_1} x_3 + \frac{b}{m_1} x_4$$

$$\dot{x}_3 = x_4$$

$$\dot{x}_4 = \frac{k_2}{m_2} x_1 + \frac{b}{m_2} x_2 - \frac{k_2}{m_2} x_3 - \frac{b}{m_2} x_4 + \frac{1}{m_2} f$$

or

$$\begin{bmatrix} \dot{x}_1 \\ \dot{x}_2 \\ \dot{x}_3 \\ \dot{x}_4 \end{bmatrix} = \begin{bmatrix} 0 & 1 & 0 & 0 \\ -\dfrac{(k_1 + k_2)}{m_1} & -\dfrac{b}{m_1} & \dfrac{k_2}{m_1} & \dfrac{b}{m_1} \\ 0 & 0 & 0 & 1 \\ \dfrac{k_2}{m_2} & \dfrac{b}{m_2} & -\dfrac{k_2}{m_2} & -\dfrac{b}{m_2} \end{bmatrix} \begin{bmatrix} x_1 \\ x_2 \\ x_3 \\ x_4 \end{bmatrix} + \begin{bmatrix} 0 \\ 0 \\ 0 \\ \dfrac{1}{m_2} \end{bmatrix} f \qquad (3)$$

The outputs are

$$y_1 = z_1 = x_1$$

$$y_2 = z_2 = x_3$$

or

$$\begin{bmatrix} y_1 \\ y_2 \end{bmatrix} = \begin{bmatrix} 1 & 0 & 0 & 0 \\ 0 & 0 & 1 & 0 \end{bmatrix} \begin{bmatrix} x_1 \\ x_2 \\ x_3 \\ x_4 \end{bmatrix} \qquad (4)$$

Equations (3) and (4) give a state-space representation of the system. By substituting the given numerical values to the state equation and defining $f = u$, we obtain

$$\begin{bmatrix} \dot{x}_1 \\ \dot{x}_2 \\ \dot{x}_3 \\ \dot{x}_4 \end{bmatrix} = \begin{bmatrix} 0 & 1 & 0 & 0 \\ -9 & -2 & 6 & 2 \\ 0 & 0 & 0 & 1 \\ 3 & 1 & -3 & -1 \end{bmatrix} \begin{bmatrix} x_1 \\ x_2 \\ x_3 \\ x_4 \end{bmatrix} + \begin{bmatrix} 0 \\ 0 \\ 0 \\ 0.05 \end{bmatrix} u$$

$$\begin{bmatrix} y_1 \\ y_2 \end{bmatrix} = \begin{bmatrix} 1 & 0 & 0 & 0 \\ 0 & 0 & 1 & 0 \end{bmatrix} \begin{bmatrix} x_1 \\ x_2 \\ x_3 \\ x_4 \end{bmatrix} + \begin{bmatrix} 0 \\ 0 \end{bmatrix} u$$

The MATLAB program presented in the next page produces the response curve $z_1(t)$, $z_2(t)$, and $z_2(t) - z_1(t)$. The response curves are shown also in the next page.

```
>> t = 0:0.02:60;
>> A = [0  1  0  0;-9  -2  6  2;0  0  0  1;3  1  -3  -1];
>> B = [0;0;0;0.05];
>> C = [1  0  0  0;0  0  1  0];
>> D = [0;0];
>> sys = ss(A,B,C,D);
>> t1 = 0:0.02:10; t2 = 10.02 :0.02:60;
>> u1 = t1; u2 = zeros(size(t2));
>> u = [u1   u2];
>> [y,t] = lsim(sys,u,t);
>> y1 = [1   0]*y';
>> y2 = [0   1]*y';
>> subplot(221); plot(t,y1); grid
>> title('Response to Triangular Input')
>> xlabel('t (sec)'); ylabel('z_1(t)')
>> subplot(222); plot(t,y2); grid
>> title('Response to Triangular Input')
>> xlabel('t (sec)'); ylabel('z_2(t)')
>> subplot(223); plot(t,y2-y1); grid
>> title('Response to Triangular Input')
>> xlabel('t (sec)'); ylabel('z_2(t) - z_1(t)')
```

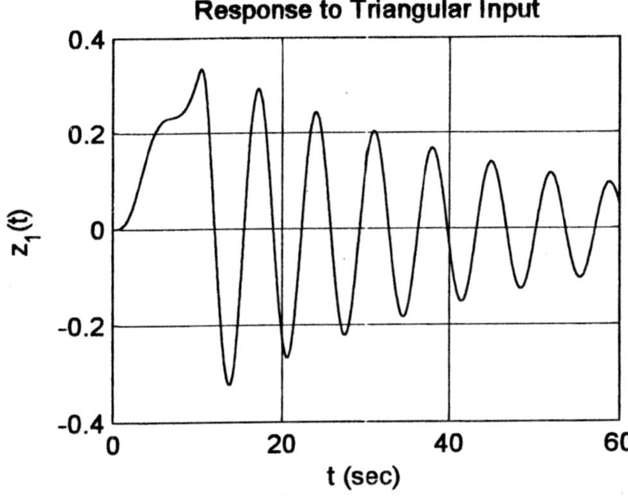

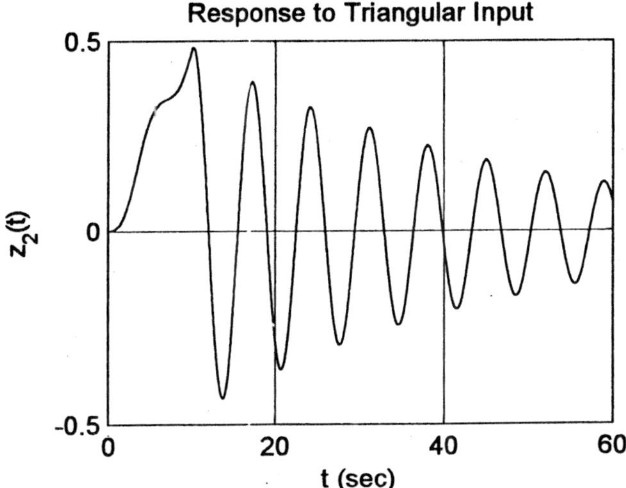

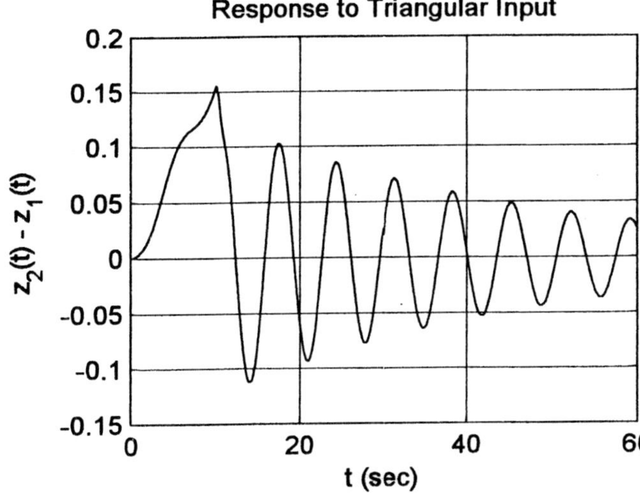

B-5-19. We shall derive Equations (5-118) and (5-119) when $n = 3$. (The derivation given here can be extended to the case where n = arbitrary positive integer.)

For $n = 3$, the given differential equation can be modified to

$$\frac{Y(s)}{U(s)} = \frac{b_0 s^3 + b_1 s^2 + b_2 s + b_3}{s^3 + a_1 s^2 + a_2 s + a_3}$$

This equation can be written as

$$s^3[Y(s) - b_0 U(s)] + s^2[a_1 Y(s) - b_1 U(s)]$$

$$+ s[a_2 Y(s) - b_2 U(s)] + a_3 Y(s) - b_3 U(s) = 0$$

By dividing the entire equation by s^3 and rearranging, we obtain

$$Y(s) = b_0 U(s) + \frac{1}{s}[b_1 U(s) - a_1 Y(s)] + \frac{1}{s^2}[b_2 U(s) - a_2 Y(s)]$$
$$+ \frac{1}{s^3}[b_3 U(s) - a_3 Y(s)]$$

or

$$Y(s) = b_0 U(s) + \frac{1}{s}\Bigg([b_1 U(s) - a_1 Y(s)] + \frac{1}{s}\Big\{[b_2 U(s) - a_2 Y(s)]$$
$$+ \frac{1}{s}[b_3 U(s) - a_3 Y(s)]\Big\}\Bigg) \tag{1}$$

Now define state variables as follows:

$$X_3(s) = \frac{1}{s}[b_1 U(s) - a_1 Y(s) + X_2(s)]$$
$$X_2(s) = \frac{1}{s}[b_2 U(s) - a_2 Y(s) + X_1(s)] \tag{2}$$
$$X_1(s) = \frac{1}{s}[b_3 U(s) - a_3 Y(s)]$$

Then, Equation (1) can be written as

$$Y(s) = b_0 U(s) + X_3(s) \tag{3}$$

By substituting Equation (3) into Equations (2) and multiplying both sides of the equations by s, we obtain

$$sX_3(s) = b_1 U(s) - a_1 Y(s) + X_2(s)$$

$$= b_1 U(s) + X_2(s) - a_1[b_0 U(s) + X_3(s)]$$

$$= X_2(s) - a_1 X_3(s) + (b_1 - a_1 b_0)U(s)$$

$$sX_2(s) = b_2 U(s) - a_2 Y(s) + X_1(s)$$

$$= X_1(s) - a_2 X_3(s) + (b_2 - a_2 b_0)U(s)$$

$$sX_1(s) = b_3U(s) - a_3Y(s)$$

$$= - a_3X_3(s) + (b_3 - a_3b_0)U(s)$$

Taking the inverse Laplace transforms of the preceding 3 equations and writing them in the reverse order, we get

$$\dot{x}_1 = - a_3x_3 + (b_3 - a_3b_0)u$$

$$\dot{x}_2 = x_1 - a_2x_3 + (b_2 - a_2b_0)u$$

$$\dot{x}_3 = x_2 - a_1x_3 + (b_1 - a_1b_0)u$$

These three equations are state equations.

Also, from Equation (3) we get

$$y = x_3 + b_0u$$

which is the output equation. Rewriting the state and output equations in the standard vector-matrix forms, we obtain Equations (5-118) and (5-119).

B-5-20. The differential equation for the system is

$$\dddot{y} + 18\ddot{y} + 192\dot{y} + 640y = 160\dot{u} + 640u$$

Method 1: By comparing this differential equation with the standard third-order differential equation:

$$\dddot{y} + a_1\ddot{y} + a_2\dot{y} + a_3y = b_0\dddot{u} + b_1\ddot{u} + b_2\dot{u} + b_3u$$

we obtain

$$a_1 = 18, \quad a_2 = 192, \quad a_3 = 640$$

$$b_0 = 0, \quad b_1 = 0, \quad b_2 = 160, \quad b_3 = 640$$

Next we compute

$$\beta_0 = b_0 = 0$$

$$\beta_1 = b_1 - a_1\beta_0 = 0$$

$$\beta_2 = b_2 - a_1\beta_1 - a_2\beta_0 = 160$$

$$\beta_3 = b_3 - a_1\beta_2 - a_2\beta_1 - a_3\beta_0 = 640 - 18 \text{ x } 160 = -2240$$

Then, the state-space equation can be obtained as

$$\begin{bmatrix} \dot{x}_1 \\ \dot{x}_2 \\ \dot{x}_3 \end{bmatrix} = \begin{bmatrix} 0 & 1 & 0 \\ 0 & 0 & 1 \\ -640 & -192 & -18 \end{bmatrix} \begin{bmatrix} x_1 \\ x_2 \\ x_3 \end{bmatrix} + \begin{bmatrix} 0 \\ 160 \\ -2240 \end{bmatrix}$$

and the output equation is

$$y = [1 \quad 0 \quad 0] \begin{bmatrix} x_1 \\ x_2 \\ x_3 \end{bmatrix}$$

Method 2: Noting that

$$a_1 = 18, \qquad a_2 = 192, \qquad a_3 = 640$$

$$b_0 = 0, \qquad b_1 = 0, \qquad b_2 = 160, \qquad b_3 = 640$$

we have

$$b_3 - a_3 b_0 = 640$$

$$b_2 - a_2 b_0 = 160$$

$$b_1 - a_1 b_0 = 0$$

Referring to Equation (5-40), we obtain the state equation as follows:

$$\begin{bmatrix} \dot{x}_1 \\ \dot{x}_2 \\ \dot{x}_3 \end{bmatrix} = \begin{bmatrix} 0 & 1 & 0 \\ 0 & 0 & 1 \\ -640 & -192 & -18 \end{bmatrix} \begin{bmatrix} x_1 \\ x_2 \\ x_3 \end{bmatrix} + \begin{bmatrix} 0 \\ 0 \\ 1 \end{bmatrix} u$$

Referring to Equation (5-41), we obtain the output equation as

$$y = [640 \quad 160 \quad 0] \begin{bmatrix} x_1 \\ x_2 \\ x_3 \end{bmatrix}$$

B-5-21.

$$\frac{Y(s)}{U(s)} = \frac{5}{(s+1)^2(s+2)}$$

The partial-fraction expansion of $Y(s)/U(s)$ is

$$\frac{Y(s)}{U(s)} = \frac{-5}{s + 1} + \frac{5}{(s + 1)^2} + \frac{5}{s + 2}$$

or

$$Y(s) = \frac{-5}{s + 1} U(s) + \frac{5}{(s + 1)^2} U(s) + \frac{5}{s + 2} U(s)$$

Define

$$\frac{-5}{s + 1} U(s) = X_1(s) \tag{1}$$

$$\frac{5}{(s + 1)^2} U(s) = X_2(s) \tag{2}$$

$$\frac{5}{s + 2} U(s) = X_3(s) \tag{3}$$

From Equation (1),we have

$$(s + 1)X_1(s) = -5U(s) \tag{4}$$

From Equations (1) and (2), we have

$$- X_1(s) = (s + 1)X_2(s) \tag{5}$$

From Equation (3), we obtain

$$(s + 2)X_3(s) = 5U(s) \tag{6}$$

Taking the inverse Laplace transforms of Equations (4), (5), and (6), we obtain

$$\dot{x}_1 + x_1 = -5u$$

$$\dot{x}_2 + x_2 + x_1 = 0$$

$$\dot{x}_3 + 2x_3 = 5u$$

The output Y(s) is

$$Y(s) = X_1(s) + X_2(s) + X_3(s)$$

Hence,

$$y = x_1 + x_2 + x_3$$

The state-equation in the form of the standard vector-matrix equation is

$$\begin{bmatrix} \dot{x}_1 \\ \dot{x}_2 \\ \dot{x}_3 \end{bmatrix} = \begin{bmatrix} -1 & 0 & 0 \\ -1 & -1 & 0 \\ 0 & 0 & -2 \end{bmatrix} \begin{bmatrix} x_1 \\ x_2 \\ x_3 \end{bmatrix} + \begin{bmatrix} -5 \\ 0 \\ 5 \end{bmatrix} u$$

The output equation is

$$y = [1 \quad 1 \quad 1] \begin{bmatrix} x_1 \\ x_2 \\ x_3 \end{bmatrix}$$

B-5-22. **The differential equation for the system is**

$$m\ddot{y} + b_1\dot{y} + (k_1 + k_2)y = u$$

This is a second-order system. Therefore, we need two state variables.

Define state variables x_1 and x_2 as follows:

$$x_1 = y$$

$$x_2 = \dot{y}$$

Then we obtain

$$\dot{x}_1 = x_2$$

$$\dot{x}_2 = -\frac{k_1 + k_2}{m} x_1 - \frac{b_1}{m} x_2 + \frac{1}{m} u$$

The output y for the system is simply x_1. Thus,

$$y = x_1$$

The state space representation for this system is given by

$$\begin{bmatrix} \dot{x}_1 \\ \dot{x}_2 \end{bmatrix} = \begin{bmatrix} 0 & 1 \\ -\frac{k_1 + k_2}{m} & -\frac{b_1}{m} \end{bmatrix} \begin{bmatrix} x_1 \\ x_2 \end{bmatrix} + \begin{bmatrix} 0 \\ \frac{1}{m} \end{bmatrix} u$$

$$y = \begin{bmatrix} 1 & 0 \end{bmatrix} \begin{bmatrix} x_1 \\ x_2 \end{bmatrix}$$

B-5-23. The equations for the system are

$$m_1\ddot{y}_1 + b\dot{y}_1 + (k_1 + k_2)y_1 = k_2y_2 + b\dot{y}_2 \tag{1}$$

$$m_2\ddot{y}_2 + b\dot{y}_2 + k_2y_2 = k_2y_1 + b\dot{y}_1 + u \tag{2}$$

The Laplace transforms of Equations (1) and (2), assuming zero initial conditions, are

$$(m_1s^2 + bs + k_1 + k_2)Y_1(s) = (k_2 + bs)Y_2(s) \tag{3}$$

$$(m_2s^2 + bs + k_2)Y_2(s) = (k_2 + bs)Y_1(s) + U(s) \tag{4}$$

From Equation (3), we have

$$Y_2(s) = \frac{m_1s^2 + bs + k_1 + k_2}{bs + k_2}\, Y_1(s) \tag{5}$$

Using Equation (5) to eliminate $Y_2(s)$ from Equation (4), we have

$$[(m_2s^2 + bs + k_2)(m_1s^2 + bs + k_1 + k_2) - (bs + k_2)^2]Y_1(s)$$
$$= (bs + k_2)U(s)$$

from which we get

$$\frac{Y_1(s)}{U(s)} = \frac{bs + k_2}{m_1m_2s^4 + (m_1 + m_2)bs^3 + [(m_1 + m_2)k_2 + m_2k_1]s^2 + bk_1s + k_1k_2} \tag{6}$$

$$= \frac{\frac{b}{m_1m_2}s + \frac{k_2}{m_1m_2}}{s^4 + \left(\frac{b}{m_1} + \frac{b}{m_2}\right)s^3 + \left(\frac{k_1}{m_1} + \frac{k_2}{m_1} + \frac{k_2}{m_2}\right)s^2 + \frac{k_1}{m_1}\frac{b}{m_2}s + \frac{k_1}{m_1}\frac{k_2}{m_2}} \tag{7}$$

From Equations (5) and (6), we obtain

$$\frac{Y_2(s)}{U(s)} = \frac{m_1s^2 + bs + k_1 + k_2}{m_1m_2s^4 + (m_1 + m_2)bs^3 + [(m_1 + m_2)k_2 + m_2k_1]s^2 + bk_1s + k_1k_2} \tag{8}$$

$$= \frac{\frac{1}{m_2}s^2 + \frac{b}{m_1m_2}s + \frac{k_1 + k_2}{m_1m_2}}{s^4 + \left(\frac{b}{m_1} + \frac{b}{m_2}\right)s^3 + \left(\frac{k_1}{m_1} + \frac{k_2}{m_1} + \frac{k_2}{m_2}\right)s_2 + \frac{k_1}{m_1}\frac{b}{m_2}s + \frac{k_1}{m_1}\frac{k_1}{m_2}} \tag{9}$$

We shall use Method 2 to obtain a state-space representation of the system. We shall first consider $Y_1(s)/U(s)$. Comparing Equation (7) with the standard 4th-order transfer function:

$$\frac{Y_1(s)}{U(s)} = \frac{b_0 s^4 + b_1 s^3 + b_2 s^2 + b_3 s + b_4}{s^4 + a_1 s^3 + a_2 s^2 + a_3 s + a_4} \qquad (10)$$

we obtain

$$a_1 = \frac{b}{m_1} + \frac{b}{m_2}, \qquad a_2 = \frac{k_1}{m_1} + \frac{k_2}{m_1} + \frac{k_2}{m_2}, \qquad a_3 = \frac{k_1 b}{m_1 m_2}, \qquad a_4 = \frac{k_1 k_2}{m_1 m_2}$$

$$b_0 = 0, \quad b_1 = 0, \quad b_2 = 0, \quad b_3 = \frac{b}{m_1 m_2}, \quad b_4 = \frac{k_2}{m_1 m_2}$$

Hence

$$b_4 - a_4 b_0 = \frac{k_2}{m_1 m_2}$$

$$b_3 - a_3 b_0 = \frac{b}{m_1 m_2}$$

$$b_2 - a_2 b_0 = 0$$

$$b_1 - a_1 b_0 = 0$$

Then, referring to Equations (5-40) and (5-41), the state equation and output equation become as follows;

$$\begin{bmatrix} \dot{x}_1 \\ \dot{x}_2 \\ \dot{x}_3 \\ \dot{x}_4 \end{bmatrix} = \begin{bmatrix} 0 & 1 & 0 & 0 \\ 0 & 0 & 1 & 0 \\ 0 & 0 & 0 & 1 \\ -\frac{k_1}{m_1}\frac{k_2}{m_2} & -\frac{k_1 b}{m_1 m_2} & -\frac{k_1}{m_1} - \frac{k_2}{m_1} - \frac{k_2}{m_2} & -\frac{b}{m_1} - \frac{b}{m_2} \end{bmatrix} \begin{bmatrix} x_1 \\ x_2 \\ x_3 \\ x_4 \end{bmatrix} + \begin{bmatrix} 0 \\ 0 \\ 0 \\ 1 \end{bmatrix} u$$

$$y_1 = \begin{bmatrix} \frac{k_2}{m_1 m_2} & \frac{b}{m_1 m_2} & 0 & 0 \end{bmatrix} \begin{bmatrix} x_1 \\ x_2 \\ x_3 \\ x_4 \end{bmatrix} + 0\,u$$

Next, we shall consider $Y_2(s)/U(s)$. Comparing Equation (9) with Equation (10), we find

$$b_0 = 0, \quad b_1 = 0, \quad b_2 = \frac{1}{m_2}, \quad b_3 = \frac{b}{m_1 m_2}, \quad b_4 = \frac{k_1 + k_2}{m_1 m_2}$$

Note that a_1, a_2, a_3, and a_4 are the same as those for $Y_1(s)/U(s)$. Hence, the state equation for the present case is the same as that for $Y_1(s)/U(s)$. Only difference is the output equation. Next, we compute

$$b_4 - a_4 b_0 = \frac{k_1 + k_2}{m_1 m_2}$$

$$b_3 - a_3 b_0 = \frac{b}{m_1 m_2}$$

$$b_2 - a_2 b_0 = \frac{1}{m_2}$$

$$b_1 - a_1 b_0 = 0$$

Hence, the output equation for $Y_2(s)/U(s)$ is

$$y_2 = \begin{bmatrix} \frac{k_1 + k_2}{m_1 m_2} & \frac{b}{m_1 m_2} & \frac{1}{m_2} & 0 \end{bmatrix} \begin{bmatrix} x_1 \\ x_2 \\ x_3 \\ x_4 \end{bmatrix} + 0\, u$$

Combining two sets of the state-space representations into one, we get

$$\begin{bmatrix} \dot{x}_1 \\ \dot{x}_2 \\ \dot{x}_3 \\ \dot{x}_4 \end{bmatrix} = \begin{bmatrix} 0 & 1 & 0 & 0 \\ 0 & 0 & 1 & 0 \\ 0 & 0 & 0 & 1 \\ -\frac{k_1 k_2}{m_1 m_2} & -\frac{k_1 b}{m_1 m_2} & -\frac{k_1}{m_1} - \frac{k_2}{m_1} - \frac{k_3}{m_2} & -\frac{b}{m_1} - \frac{b}{m_2} \end{bmatrix} \begin{bmatrix} x_1 \\ x_2 \\ x_3 \\ x_4 \end{bmatrix} + \begin{bmatrix} 0 \\ 0 \\ 0 \\ 1 \end{bmatrix} u \qquad (11)$$

$$\begin{bmatrix} y_1 \\ y_2 \end{bmatrix} = \begin{bmatrix} \frac{k_2}{m_1 m_2} & \frac{b}{m_1 m_2} & 0 & 0 \\ \frac{k_1 + k_2}{m_1 m_2} & \frac{b}{m_1 m_2} & \frac{1}{m_2} & 0 \end{bmatrix} \begin{bmatrix} x_1 \\ x_2 \\ x_3 \\ x_4 \end{bmatrix} + \begin{bmatrix} 0 \\ 0 \end{bmatrix} u \qquad (12)$$

Substituting $m_1 = 10$, $m_2 = 5$, $b = 10$, $k_1 = 40$, $k_2 = 20$ and $u = 5 \cdot 1(t)$ into Equations (11) and (12), we get

$$\begin{bmatrix} \dot{x}_1 \\ \dot{x}_2 \\ \dot{x}_3 \\ \dot{x}_4 \end{bmatrix} = \begin{bmatrix} 0 & 1 & 0 & 0 \\ 0 & 0 & 1 & 0 \\ 0 & 0 & 0 & 1 \\ -16 & -3 & -10 & -3 \end{bmatrix} \begin{bmatrix} x_1 \\ x_2 \\ x_3 \\ x_4 \end{bmatrix} + \begin{bmatrix} 0 \\ 0 \\ 0 \\ 5 \end{bmatrix} 1(t)$$

$$\begin{bmatrix} y_1 \\ y_2 \end{bmatrix} = \begin{bmatrix} 0.4 & 0.2 & 0 & 0 \\ 1.2 & 0.2 & 0.2 & 0 \end{bmatrix} \begin{bmatrix} x_1 \\ x_2 \\ x_3 \\ x_4 \end{bmatrix} + \begin{bmatrix} 0 \\ 0 \end{bmatrix} 1(t)$$

A MATLAB program to plot $y_1(t)$ versus t and $y_2(t)$ versus t is given next. The resulting plots are shown on the next page.

```
>> A = [0  1  0  0;0  0  1  0;0  0  0  1;-16  -8  -10  -3];
>> B = [0;0;0;5];
>> C = [0.4  0.2  0  0;1.2  0.2  0.2  0];
>> D = [0;0];
>> t = 0:0.01:30;
>> y = step(A,B,C,D,1,t);
>> y1 = [1   0]*y'; y2 = [0   1]*y';
>> subplot(221); plot(t,y1)
>> grid;
>> title('Step Response y_1 versus t when u = 5*1(t)')
>> xlabel('t (sec)')
>> ylabel('Response y_1')
>> subplot(222); plot(t,y2)
>> grid
>> title('Step Response y_2 versus t when u = 5*1(t)')
>> xlabel('t (sec)')
>> ylabel('Response y_2')
```

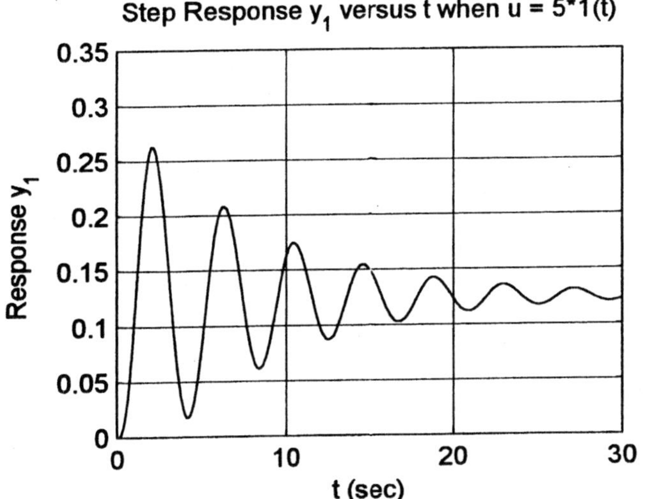

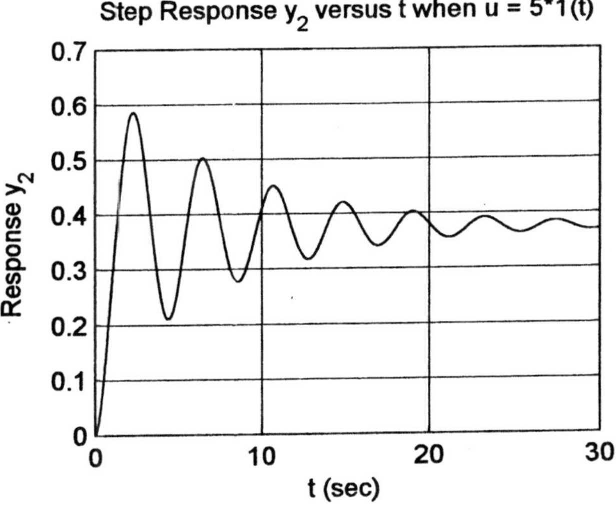

B-5-24. The equations for the system are

$$k_1(u - z) = b_1(\dot{z} - \dot{y})$$

$$b_1(\dot{z} - \dot{y}) = k_2 y$$

or

$$k_1 u + b_1 \dot{y} = b_1 \dot{z} + k_1 z$$

$$b_1 \dot{z} = b_1 \dot{y} + k_2 y$$

By taking Laplace transforms of these two equations, assuming the zero initial conditions, we obtain

$$(b_1 s + k_1)Z(s) = b_1 sY(s) + k_1 U(s)$$

$$(b_1 s + k_2)Y(s) = b_1 sZ(s) \qquad (1)$$

By eliminating Z(s) from the last two equations,

$$(b_1 s + k_1)\,\frac{b_1 s + k_2}{b_1 s}\,Y(s) = b_1 sY(s) + k_1 U(s)$$

or

$$(b_1 s + k_1)(b_1 s + k_2)Y(s) = b_1^2\, s^2 Y(s) + b_1 s k_1 U(s)$$

which can be simplified to

$$[(k_1 + k_2)b_1 s + k_1 k_2]Y(s) = k_1 b_1 sU(s)$$

Hence

$$\frac{Y(s)}{U(s)} = \frac{k_1 b_1 s}{(k_1 + k_2)b_1 s + k_1 k_2} = \frac{\frac{b_1}{k_2} s}{\frac{k_1 + k_2}{k_1 k_2} b_1 s + 1} \tag{2}$$

Using Equations (1) and (2), Z(s)/U(s) can be obtained as follows:

$$\frac{Z(s)}{U(s)} = \frac{Z(s)}{Y(s)} \frac{Y(s)}{U(s)} = \frac{b_1 s + k_2}{b_1 s} \frac{k_1 b_1 s}{(k_1 + k_2)b_1 s + k_1 k_2}$$

$$= \frac{b_1 k_1 s + k_1 k_2}{(k_1 + k_2)b_1 s + k_1 k_2}$$

$$= \frac{\frac{b_1}{k_2} s + 1}{\frac{k_1 + k_2}{k_1 k_2} b_1 s + 1} \tag{3}$$

Define

$$\frac{b_1}{k_2} = T, \qquad \frac{k_1 + k_2}{k_1 k_2} b_1 = aT$$

where

$$a = \frac{K_1 + k_2}{k_1}$$

Then, Equations (2) and (3) can be written as

$$\frac{Y(s)}{U(s)} = \frac{Ts}{aTs + 1}, \qquad \frac{Z(s)}{U(s)} = \frac{Ts + 1}{aTs + 1}$$

Hence

$$aTsY(s) - TsU(s) = -Y(s) \tag{4}$$

$$aTsZ(s) - TsU(s) = -Z(s) + U(s) \tag{5}$$

Define

$$aY(s) - U(s) = X_1(s)$$

$$aZ(s) - U(s) = X_2(s)$$

Then

$$Y(s) = \frac{1}{a} [X_1(s) + U(s)] \tag{6}$$

$$Z(s) = \frac{1}{a} [X_2(s) + U(s)] \tag{7}$$

Equations (4) and (5) can be written as

$$TsX_1(s) = -Y(s) = -\frac{1}{a}[X_1(s) + U(s)]$$

$$TsX_2(s) = -Z(s) + U(s) = -\frac{1}{a}[X_2(s) + U(s)] + U(s)$$

Rewriting,

$$TsX_1(s) = -\frac{1}{a}X_1(s) - \frac{1}{a}U(s)$$

$$TsX_2(s) = -\frac{1}{a}X_2(s) - \frac{1}{a}U(s) + U(s)$$

from which we get

$$\dot{x}_1 = -\frac{1}{aT}x_1 - \frac{1}{aT}u$$

$$\dot{x}_2 = -\frac{1}{aT}x_2 + \frac{a-1}{aT}u$$

Hence

$$\begin{bmatrix} \dot{x}_1 \\ \dot{x}_2 \end{bmatrix} = \begin{bmatrix} -\frac{1}{aT} & 0 \\ 0 & -\frac{1}{aT} \end{bmatrix} \begin{bmatrix} x_1 \\ x_2 \end{bmatrix} + \begin{bmatrix} -\frac{1}{aT} \\ \frac{a-1}{aT} \end{bmatrix} u$$

This is the state equation.

From Equation (6) we have

$$y = \frac{1}{a}x_1 + \frac{1}{a}u$$

From Equation (7) we get

$$z = \frac{1}{a}x_2 + \frac{1}{a}u$$

Hence

$$\begin{bmatrix} y \\ z \end{bmatrix} = \begin{bmatrix} \frac{1}{a} & 0 \\ 0 & \frac{1}{a} \end{bmatrix} \begin{bmatrix} x_1 \\ x_2 \end{bmatrix} + \begin{bmatrix} \frac{1}{a} \\ \frac{1}{a} \end{bmatrix} u$$

This is the output equation.

B-5-25. Define the displacement of the point where the force u is applied as y. (See the figure below.)

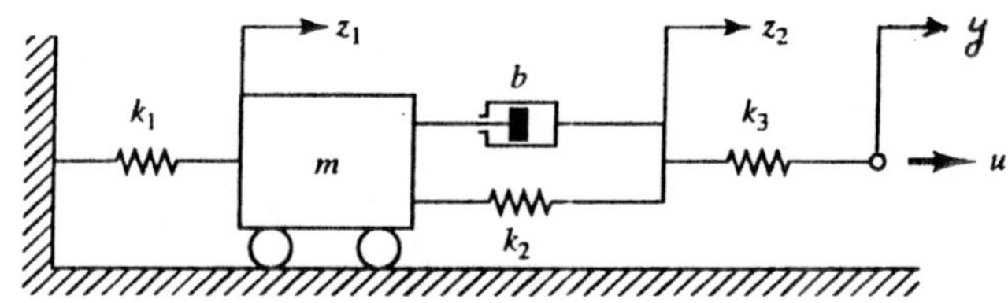

The equations for the system are

$$m\ddot{z}_1 = -k_1 z_1 - b(\dot{z}_1 - \dot{z}_2) - k_2(z_1 - z_2)$$

$$b(\dot{z}_2 - \dot{z}_1) + k_2(z_2 - z_1) = k_3(y - z_2) = u$$

Rewriting these equations, we obtain

$$m\ddot{z}_1 + b\dot{z}_1 + (k_1 + k_2)z_1 = b\dot{z}_2 + k_2 z_2 \tag{1}$$

$$b\dot{z}_2 + k_2 z_2 = b\dot{z}_1 + k_2 z_1 + u \tag{2}$$

Taking Laplace transforms of Equations (1) and (2), assuming zero initial conditions, we get

$$(ms^2 + bs + k_1 + k_2)Z_1(s) = (bs + k_2)Z_2(s) \tag{3}$$

$$(bs + k_2)Z_2(s) = (bs + k_2)Z_1(s) + U(s) \tag{4}$$

By eliminating $z_2(s)$ from Equations (3) and (4), we get

$$(ms^2 + bs + k_1 + k_2)Z_1(s) = (bs + k_2)Z_1(s) + U(s)$$

or

$$(ms^2 + k_1)Z_1(s) = U(s)$$

Hence

$$\frac{Z_1(s)}{U(s)} = \frac{1}{ms^2 + k_1} \tag{5}$$

From Equation (4), we have

$$Z_2(s) = Z_1(s) + \frac{1}{bs + k_2} U(s) \tag{6}$$

Define

$$z = z_2 - z_1$$

Then, from Equation (6) we get

$$Z(s) = \frac{1}{bs + k_2} U(s)$$

or

$$b\dot{z} + k_2 z = u \qquad (7)$$

Define also

$$x_3 = z$$

Then, from Equation (7) we get

$$\dot{x}_3 = -\frac{k_2}{b} x_3 + \frac{1}{b} u \qquad (8)$$

From Equation (5),

$$(ms^2 + k_1)Z_1(s) = U(s)$$

or

$$m\ddot{z}_1 + k_1 z_1 = u \qquad (9)$$

Next, define state variables for the system of Equation (9) as follows:

$$x_1 = z_1$$

$$x_2 = \dot{z}_1$$

Then, we have

$$\dot{x}_1 = x_2 \qquad (10)$$

From Equation (9), we get

$$\dot{x}_2 = -\frac{k_1}{m} x_1 + \frac{1}{m} u \qquad (11)$$

From Equations (8), (10), and (11), we have

$$\begin{bmatrix} \dot{x}_1 \\ \dot{x}_2 \\ \dot{x}_3 \end{bmatrix} = \begin{bmatrix} 0 & 1 & 0 \\ -\frac{k_1}{m} & 0 & 0 \\ 0 & 0 & -\frac{k_2}{b} \end{bmatrix} \begin{bmatrix} x_1 \\ x_2 \\ x_3 \end{bmatrix} + \begin{bmatrix} 0 \\ \frac{1}{m} \\ \frac{1}{b} \end{bmatrix} u \qquad (12)$$

The outputs of the system are z_1 and z_2. Since $z_1 = x_1$ and $z_2 = z_1 + z = x_1 + x_3$, the output equation is

$$\begin{bmatrix} z_1 \\ z_2 \end{bmatrix} = \begin{bmatrix} 1 & 0 & 0 \\ 1 & 0 & 1 \end{bmatrix} \begin{bmatrix} x_1 \\ x_2 \\ x_3 \end{bmatrix} + \begin{bmatrix} 0 \\ 0 \end{bmatrix} u \qquad (13)$$

Equations (12) and (13) are a state equation and an output equation of the system, respectively.

<u>B-5-26</u>. The equations for the system are

$$m_1\ddot{z}_1 = - b\dot{z}_1 - k_1z_1 - k_2(z_1 - z_2)$$

$$m_2\ddot{z}_2 = - k_2(z_2 - z_1) + u$$

Rewriting these equations we get

$$m_1\ddot{z}_1 + b\dot{z}_1 + k_1z_1 + k_2z_1 = k_2z_2 \qquad (1)$$

$$m_2\ddot{z}_2 + k_2z_2 = k_2z_1 + u \qquad (2)$$

Laplace transforming Equations (1) and (2), assuming zero initial conditions, we get

$$(m_1s^2 + bs + k_1 + k_2)Z_1(s) = k_2Z_2(s) \qquad (3)$$

$$(m_2s^2 + k_2)Z_2(s) = k_2Z_1(s) + U(s) \qquad (4)$$

From Equation (3), we have

$$Z_2(s) = \frac{m_1s^2 + bs + k_1 + k_2}{k_2} Z_1(s) \qquad (5)$$

Substituting Equation (5) into Equation (4), we get

$$\frac{(m_2s^2 + k_2)(m_1s^2 + bs + k_1 + k_2)}{k_2} Z_1(s) = k_2Z_1(s) + U(s)$$

or

$$\frac{Z_1(s)}{U(s)} = \frac{k_2}{m_1m_2s^4 + bm_2s^3 + (k_2m_1 + m_2k_1 + m_2k_2)s^2 + bk_2s + k_1K_2}$$

$$= \frac{\dfrac{k_2}{m_1 m_2}}{s^4 + \dfrac{b}{m_1} s^3 + \left(\dfrac{k_2}{m_2} + \dfrac{k_1}{m_1} + \dfrac{k_2}{m_1}\right) s^2 + \dfrac{b\,k_2}{m_1 m_2} s + \dfrac{k_1 k_2}{m_1 m_2}} \tag{6}$$

Comparing this transfer function with the standard 4th-order transfer function:

$$\frac{Z_1(s)}{U(s)} = \frac{b_0 s^4 + b_1 s^3 + b_2 s^2 + b_3 s + b_4}{s^4 + a_1 s^3 + a_2 s^2 + a_3 s + a_4}$$

we obtain

$$a_1 = \frac{b}{m_1}, \qquad a_2 = \frac{k_2}{m_2} + \frac{k_1}{m_1} + \frac{k_2}{m_1}, \qquad a_3 = \frac{bk_2}{m_1 m_2}, \qquad a_4 = \frac{k_1 k_2}{m_1 m_2}$$

$$b_0 = 0, \qquad b_1 = 0, \qquad b_2 = 0, \qquad b_3 = 0, \qquad b_4 = \frac{k_2}{m_1 m_2}$$

Using Method 2 and referring to Equations (5-40) and (5-41), we get the state-space representation for the system defined by Equation (6) as follows:

$$\begin{bmatrix} \dot{x}_1 \\ \dot{x}_2 \\ \dot{x}_3 \\ \dot{x}_4 \end{bmatrix} = \begin{bmatrix} 0 & 1 & 0 & 0 \\ 0 & 0 & 1 & 0 \\ 0 & 0 & 0 & 1 \\ -a_4 & -a_3 & -a_2 & -a_1 \end{bmatrix} \begin{bmatrix} x_1 \\ x_2 \\ x_3 \\ x_4 \end{bmatrix} + \begin{bmatrix} 0 \\ 0 \\ 0 \\ 1 \end{bmatrix} u \tag{7}$$

$$z_1 = \begin{bmatrix} \dfrac{k_2}{m_1 m_2} & 0 & 0 & 0 \end{bmatrix} \begin{bmatrix} x_1 \\ x_2 \\ x_3 \\ x_4 \end{bmatrix} + \begin{bmatrix} 0 \\ 0 \end{bmatrix} u \tag{8}$$

Next consider $Z_2(s)/U(s)$:

$$\frac{Z_2(s)}{U(s)} = \frac{Z_1(s)}{U(s)} \frac{Z_2(s)}{Z_1(s)}$$

$$= \frac{m_1 s^2 + bs + k_1 + k_2}{m_1 m_2 s^4 + bm_2 s^3 + (k_2 m_1 + m_2 k_1 + m_2 k_2) s^2 + bk_2 s + k_1 k_2}$$

$$= \frac{\dfrac{1}{m_2} s^2 + \dfrac{b}{m_1 m_2} s + \dfrac{k_1 + k_2}{m_1 m_2}}{s^4 + \dfrac{b}{m_1} s^3 + \left(\dfrac{k_2}{m_2} + \dfrac{k_1}{m_1} + \dfrac{k_2}{m_1}\right) s^2 + \dfrac{b\,k_2}{m_1 m_2} s + \dfrac{k_1 k_2}{m_1 m_2}} \tag{9}$$

For this case

$$b_0 = 0, \quad b_1 = 0, \quad b_2 = \frac{1}{m_2}, \quad b_3 = \frac{b}{m_1 m_2}, \quad b_4 = \frac{k_1 + k_2}{m_1 m_2}$$

Hence

$$b_4 - a_4 b_0 = \frac{k_1 + k_2}{m_1 m_2}$$

$$b_3 - a_3 b_0 = \frac{b}{m_1 m_2}$$

$$b_2 - a_2 b_0 = \frac{1}{m_2}$$

$$b_1 - a_1 b_0 = 0$$

Referring to Equation (5-40), the state equation for the system of Equation (9) is the same as Equation (7), but the output equation is different from Equation (8) and is given by

$$z_2 = \begin{bmatrix} \frac{k_1 + k_2}{m_1 m_2} & \frac{b}{m_1 m_2} & \frac{1}{m_2} & 0 \end{bmatrix} \begin{bmatrix} x_1 \\ x_2 \\ x_3 \\ x_4 \end{bmatrix} + 0\,u \qquad (10)$$

Combining two sets of state-space representations [one given by Equations (7) and (8) and the other given by Equations (7) and (10)], we get one state-space representation for the entire system as follows:

$$\begin{bmatrix} \dot{x}_1 \\ \dot{x}_2 \\ \dot{x}_3 \\ \dot{x}_4 \end{bmatrix} = \begin{bmatrix} 0 & 1 & 0 & 0 \\ 0 & 0 & 1 & 0 \\ 0 & 0 & 0 & 1 \\ -\frac{k_1 k_2}{m_1 m_2} & -\frac{b k_2}{m_1 m_2} & -\frac{k_2}{m_2} - \frac{k_1}{m_1} - \frac{k_2}{m_1} & -\frac{b}{m_1} \end{bmatrix} \begin{bmatrix} x_1 \\ x_2 \\ x_3 \\ x_4 \end{bmatrix} + \begin{bmatrix} 0 \\ 0 \\ 0 \\ 1 \end{bmatrix} u \qquad (11)$$

$$\begin{bmatrix} z_1 \\ z_2 \end{bmatrix} = \begin{bmatrix} \frac{k_2}{m_1 m_2} & 0 & 0 & 0 \\ \frac{k_1 + k_2}{m_1 m_2} & \frac{b}{m_1 m_2} & \frac{1}{m_2} & 0 \end{bmatrix} \begin{bmatrix} x_1 \\ x_2 \\ x_3 \\ x_4 \end{bmatrix} + \begin{bmatrix} 0 \\ 0 \end{bmatrix} u \qquad (12)$$

Equations (11) and (12) are the state-space representation of the system based on Method 2.

For the given numerical values $m_1 = 100$, $m_2 = 200$, $b = 25$, $k_1 = 50$, and $k_2 = 100$, we have

$$\frac{k_1 k_2}{m_1 m_2} = 0.25$$

$$\frac{b k_2}{m_1 m_2} = 0.125$$

$$\frac{k_2}{m_2} + \frac{k_1}{m_1} + \frac{k_2}{m_1} = 2$$

$$\frac{b}{m_1} = 0.25$$

$$\frac{k_2}{m_1 m_2} = 0.005$$

$$\frac{k_1 + k_2}{m_1 m_2} = 0.0075$$

$$\frac{b}{m_1 m_2} = 0.00125$$

$$\frac{1}{m_2} = 0.005$$

Using these numbers, we can obtain the state-space representation of the system as follows [Note that $u = 10 \cdot 1(t)$.]:

$$\begin{bmatrix} \dot{x}_1 \\ \dot{x}_2 \\ \dot{x}_3 \\ \dot{x}_4 \end{bmatrix} = \begin{bmatrix} 0 & 1 & 0 & 0 \\ 0 & 0 & 1 & 0 \\ 0 & 0 & 0 & 1 \\ -0.25 & -0.125 & -2 & 0.25 \end{bmatrix} \begin{bmatrix} x_1 \\ x_2 \\ x_3 \\ x_4 \end{bmatrix} + \begin{bmatrix} 0 \\ 0 \\ 0 \\ 1 \end{bmatrix} 10 \cdot 1(t)$$

$$\begin{bmatrix} z_1 \\ z_2 \end{bmatrix} = \begin{bmatrix} 0.005 & 0 & 0 & 0 \\ 0.0075 & 0.00125 & 0.005 & 0 \end{bmatrix} \begin{bmatrix} x_1 \\ x_2 \\ x_3 \\ x_4 \end{bmatrix} + \begin{bmatrix} 0 \\ 0 \end{bmatrix} 10 \cdot 1(t)$$

MATLAB program to obtain the step responses $z_1(t)$ and $z_2(t)$ to the input $u = 10 \cdot 1(t)$ is given on the next page.

```
>> A = [0  1  0  0;0  0  1  0;0  0  0  1;-0.25  -0.125  -2  -0.25];
>> B = [0;0;0;10];
>> C = [0.005  0  0  0;0.0075  0.00125  0.005  0];
>> D = [0;0];
>> t = 0:0.05:200;
>> [z,x,t] = step(A,B,C,D,1,t);
>> z1 = [1  0]*z';
>> z2 = [0  1]*z';
>> subplot(221); plot(t,z1)
>> grid
>> title('Step Response')
>> xlabel('t (sec)')
>> ylabel('z_1(t)')
>> subplot(222); plot(t,z2)
>> grid
>> title('Step Response')
>> xlabel('t (sec)')
>> ylabel('z_2(t)')
>> subplot(223); plot(t,z2-z1)
>> grid
>> title('Step Response')
>> xlabel('t (sec)')
>> ylabel('z_2(t)- z_1(t)')
```

The resulting response curves $z_1(t)$ versus t and $z_2(t)$ versus t are shown below. The response curve $z_2(t) - z_1(t)$ versus t is shown on the next page.

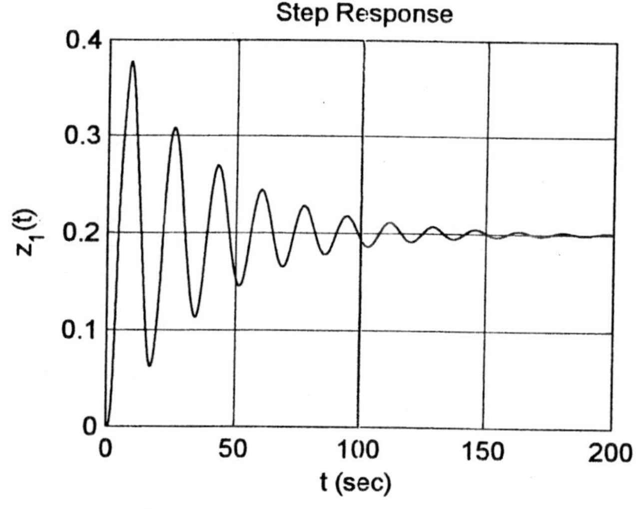

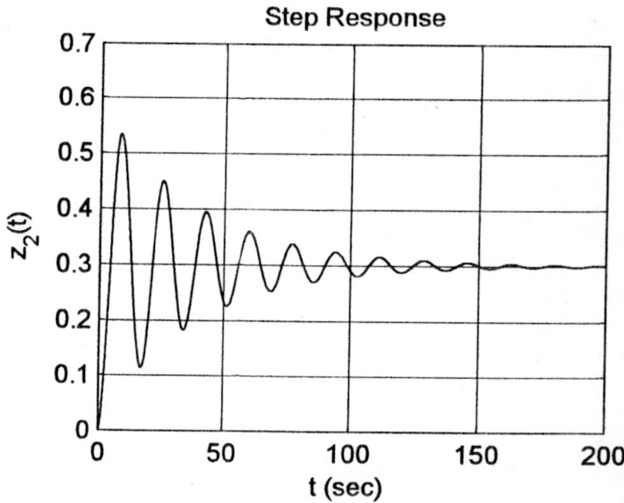

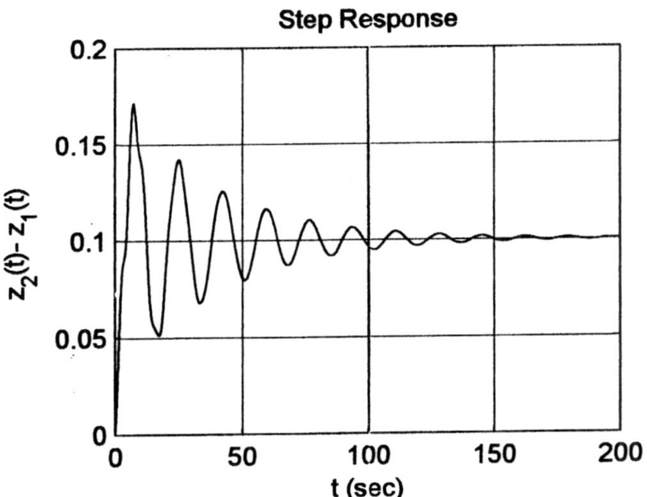

B-5-27. A possible MATLAB program to solve the problem is given below.

```
>> num = [25   5];
>> den = [1   5   25   5];
>> [A,B,C,D] = tf2ss(num,den)

A =

    -5   -25    -5
     1     0     0
     0     1     0


B =

     1
     0
     0


C =

     0    25     5


D =

     0
```

B-5-28. A possible MATLAB program to solve the problem is shown below.

```
>> num = [1    2    15    10];
>> den = [1    4    8    10];
>> [A,B,C,D] = tf2ss(num,den)

A =

    -4    -8   -10
     1     0     0
     0     1     0


B =

     1
     0
     0


C =

    -2     7     0


D =

     1
```

B-5-29. The transfer matrix G for the system obtained by use of MATLAB program presented on the next page is shown below.

$$\underset{\sim}{G} = \begin{bmatrix} \dfrac{s^2 + 5s + 29}{s^2 + 4s + 25} & \dfrac{s + 5}{s^2 + 4s + 25} \\ \\ \dfrac{-25}{s^2 + 4s + 25} & \dfrac{s^2 + 5s}{s^2 + 4s + 25} \end{bmatrix}$$

Note that to obtain the transfer matrix $\underset{\sim}{G}$, we use the following MATLAB commands:

```
[Num,denI = ss2tf(A,B,C,D,1)

[Num,den] = ss2tf(A,B,C,D,2)
```

```
>> A = [0   1;-25   -4];
>> B = [1   1;0   1];
>> C = [1   0;0   1];
>> D = [1   0;0   1];
>> [Num,den] = ss2tf(A,B,C,D,1)

Num =

    1.0000    5.0000   29.0000
         0         0  -25.0000


den =

    1.0000    4.0000   25.0000

>> [Num,den] = ss2tf(A,B,C,D,2)

Num =

         0    1.0000    5.0000
    1.0000    5.0000         0


den =

    1.0000    4.0000   25.0000
```

B-5-30. The transfer matrix $\underset{\sim}{G}$ of the system obtained by use of the MATLAB program shown on the next page is as follows:

$$\underset{\sim}{G} = \begin{bmatrix} \dfrac{1}{s^3 + 6s^2 + 4s + 2} & \dfrac{s + 6}{s^3 + 6s^2 + 4s + 2} \\ \\ \dfrac{s}{s^3 + 6s^2 + 4s + 2} & \dfrac{s^2 + 6s}{s^3 + 6s^2 + 4s + 2} \end{bmatrix}$$

```
>> A = [0   1   0;0   0   1;-2   -4   -6];
>> B = [0   0;0   1;1   0];
>> C = [1   0   0;0   1   0];
>> D = [0   0;0   0];
>> [Num,den] = ss2tf(A,B,C,D,1)

Num =

          0         0   -0.0000    1.0000
          0   -0.0000    1.0000   -0.0000


den =

    1.0000    6.0000    4.0000    2.0000

>> [Num,den] = ss2tf(A,B,C,D,2)

Num =

          0   -0.0000    1.0000    6.0000
          0    1.0000    6.0000    0.0000


den =

    1.0000    6.0000    4.0000    2.0000
```

B-5-31. Define

$$\mathbf{B} = \begin{bmatrix} \lambda_1 & 1 & 0 \\ 0 & \lambda_1 & 0 \\ 0 & 0 & \lambda_1 \end{bmatrix}$$

The eigenvectors for this matrix can be determined by solving the following equation for $\mathbf{x}$.

$$\mathbf{B}\mathbf{x} = \lambda_1 \mathbf{x}$$

or

$$\begin{bmatrix} \lambda_1 & 1 & 0 \\ 0 & \lambda_1 & 0 \\ 0 & 0 & \lambda_1 \end{bmatrix} \begin{bmatrix} x_1 \\ x_2 \\ x_3 \end{bmatrix} = \lambda_1 \begin{bmatrix} x_1 \\ x_2 \\ x_3 \end{bmatrix}$$

which can be rewritten as

$$\lambda_1 x_1 + x_2 = \lambda_1 x_1$$

$$\lambda_1 x_2 = \lambda_1 x_2$$

$$\lambda_1 x_3 = \lambda_1 x_3$$

which, in turn, given

$$x_1 = \text{arbitrary constant}$$

$$x_2 = 0$$

$$x_3 = \text{arbitray constant}$$

Hence

$$\underset{\sim}{x} = \begin{bmatrix} a \\ 0 \\ b \end{bmatrix}$$

where a and b are arbitrary nonzero constant. Notice that

$$\begin{bmatrix} a \\ 0 \\ b \end{bmatrix} = a \begin{bmatrix} 1 \\ 0 \\ 0 \end{bmatrix} + b \begin{bmatrix} 0 \\ 0 \\ 1 \end{bmatrix}$$

Note that

$$\begin{bmatrix} 1 \\ 0 \\ 0 \end{bmatrix}, \quad \begin{bmatrix} 0 \\ 1 \\ 0 \end{bmatrix}, \quad \begin{bmatrix} 0 \\ 0 \\ 1 \end{bmatrix}$$

are linearly independent vectors. Thus, the eigenvectors of matrix $\underset{\sim}{B}$ involve two linearly independent eigenvectors.

Next, define

$$\underset{\sim}{C} = \begin{bmatrix} \lambda_1 & 0 & 0 \\ 0 & \lambda_1 & 0 \\ 0 & 0 & \lambda_1 \end{bmatrix}$$

The eigenvectors for this matrix can be determined by solving the following equation for $\underset{\sim}{x}$.

$$\underset{\sim}{C}\underset{\sim}{x} = \lambda_1 \underset{\sim}{x}$$

or

$$\begin{bmatrix} \lambda_1 & 0 & 0 \\ 0 & \lambda_1 & 0 \\ 0 & 0 & \lambda_1 \end{bmatrix} \begin{bmatrix} x_1 \\ x_2 \\ x_3 \end{bmatrix} = \lambda_1 \begin{bmatrix} x_1 \\ x_2 \\ x_3 \end{bmatrix}$$

which can be rewritten as

$$\lambda_1 x_1 = \lambda_1 x_1$$

$$\lambda_1 x_2 = \lambda_1 x_2$$

$$\lambda_1 x_3 = \lambda_1 x_3$$

from which we obtain

$$x_1 = \text{arbitrary constant}$$

$$x_2 = \text{arbitrary constant}$$

$$x_3 = \text{arbitrary constant}$$

Hence

$$\underset{\sim}{x} = \begin{bmatrix} a \\ b \\ c \end{bmatrix}$$

where a, b, and c are arbitrary nonzero constant. Notice that

$$\begin{bmatrix} a \\ b \\ c \end{bmatrix} = a \begin{bmatrix} 1 \\ 0 \\ 0 \end{bmatrix} + b \begin{bmatrix} 0 \\ 1 \\ 0 \end{bmatrix} + c \begin{bmatrix} 0 \\ 0 \\ 1 \end{bmatrix}$$

Thus, the eigenvectors of matrix $\underset{\sim}{C}$ involve three linearly independent eigenvectors.

CHAPTER 6

B-6-1.

$$\frac{1}{R_{AB}} = \frac{1}{R_1} + \frac{1}{R_2 + R_3} = \frac{R_2 + R_3 + R_1}{R_1(R_2 + R_3)}$$

$$R_{AB} = \frac{R_1(R_2 + R_3)}{R_1 + R_2 + R_3}$$

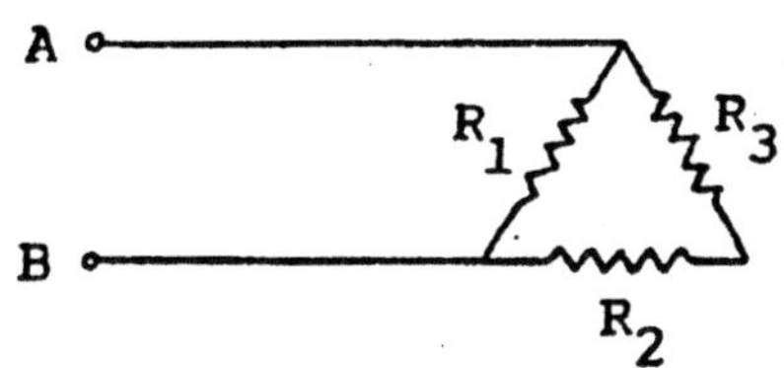

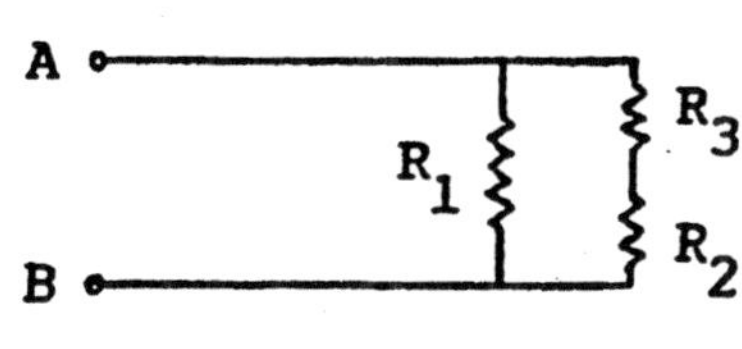

B-6-2. Figure 6-68 can be redrawn as shown below.

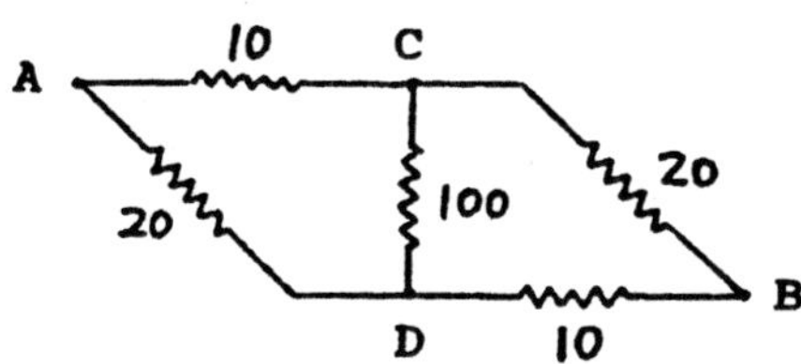

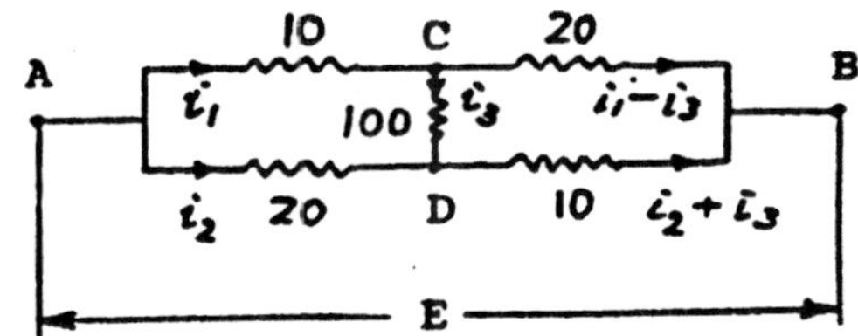

From the right side diagram we obtain the following equations:

$$10\, i_1 + 20(i_1 - i_3) = E$$

$$20\, i_2 + 10(i_2 + i_3) = E$$

$$100\, i_3 + 10(i_2 + i_3) = 20(i_1 - i_3)$$

which yield

$$i_1 = \frac{12}{11} i_2, \qquad i_3 = \frac{1}{11} i_2$$

So we obtain

$$10\, i_1 + 20(i_1 - i_3) = \frac{120}{11} i_2 + 20(\frac{12}{11} - \frac{1}{11})i_2 = E$$

Solving for i_2, we get

$$i_2 = \frac{11}{340} E$$

Thus,

$$R_{AB} = \frac{E}{i_1 + i_2} = \frac{E}{\left(\frac{12}{11} + 1\right) i_2} = \frac{E}{\frac{23}{11} \times \frac{11}{340} E} = \frac{340}{23} = 14.78\ \Omega$$

B-6-3. When switch S is open, resistance between points A and B is $160\ \Omega$ and we have

$$i_0 = \frac{E}{160}$$

When switch S is closed, resistance between points A and B is

$$60 + \frac{1}{\frac{1}{100} + \frac{1}{R}} = 60 + \frac{100\ R}{100 + R}$$

So we have

$$2i_0 = \frac{E}{60 + \frac{100\ R}{100 + R}}$$

Consequently,

$$\frac{2E}{160} = \frac{E}{60 + \frac{100\ R}{100 + R}}$$

Solving for R, we obtain

$$R = 25\ \Omega$$

B-6-4. From the circuit diagram we obtain

$$L \frac{di_1}{dt} + R_1 i_1 + R_2 (i_1 - i_2) = e$$

$$R_3 i_2 + \frac{1}{C_2} \int i_2\, dt + R_2 (i_2 - i_1) = 0$$

or

$$L \frac{di_1}{dt} + (R_1 + R_2)i_1 - R_2 i_2 = e$$

$$- R_2 i_1 + (R_2 + R_3)i_2 + \frac{1}{C_2} \int i_2 \, dt = 0$$

Each of the preceding two sets of equations constitutes a mathematical model for the circuit.

B-6-5. From Figure 6-71, we have for $t > 0$

$$R_1 i_1 + R_2(i_1 - i_2) = 0 \qquad (1)$$

$$\frac{1}{C} \int i_2 \, dt + R_2(i_2 - i_1) = 0 \qquad (2)$$

Taking Laplace transforms of Equations (1) and (2), we get

$$R_1 I_1(s) + R_2[I_1(s) - I_2(s)] = 0 \qquad (3)$$

$$\frac{1}{C}\left[\frac{I_2(s)}{s} + \frac{i_2^{-1}(0)}{s}\right] + R_2[I_2(s) - I_1(s)] = 0 \qquad (4)$$

where $i_2^{-1}(0)$ is given by

$$i_2^{-1}(0) = \int i_2(t) \, dt \Big|_{t = 0}$$

$$= q(0) = e_0 C$$

From Equation (3) we have

$$I_2(s) = \frac{R_1 + R_2}{R_2} I_1(s) \qquad (5)$$

Then Equation (4) can be rewritten as

$$\left(\frac{1}{Cs} + R_2\right) I_2(s) = R_2 I_1(s) - \frac{e_0}{s} \qquad (6)$$

Substituting Equation (5) into Equation (6) we obtain

$$\left(\frac{1}{Cs} + R_2\right) \frac{R_1 + R_2}{R_2} I_1(s) = R_2 I_1(s) - \frac{e_0}{s}$$

or

$$I_1(s) = -\frac{e_0 R_2 C}{R_1 + R_2 + R_1 R_2 C s}$$

$$= -\frac{e_0}{R_1} \frac{1}{s + \dfrac{R_1 + R_2}{R_1 R_2 C}}$$

The inverse Laplace transform of this last equation gives

$$i_1(t) = -\frac{e_0}{R_1} \exp[-(R_1 + R_2)t/(R_1 R_2 C)] \qquad t > 0$$

Referring to Equation (5) we have

$$i_2(t) = -\frac{(R_1 + R_2)e_0}{R_1 R_2} \exp[-(R_1 + R_2)t/(R_1 R_2 C)] \qquad t > 0$$

B-6-6. At steady state ($t < 0$) we have

$$i_0 = \frac{E}{R_1 + R_2}$$

and

$$e_C = \frac{1}{C}\int i\,dt = R_2 i_0 = \frac{R_2}{R_1 + R_2} E$$

For $t \geqslant 0$ the equation for the circuit is

$$R_1 i + \frac{1}{C}\int i\,dt = E$$

By differentiating this last equation, we obtain

$$R_1 \frac{di}{dt} + \frac{1}{C} i = 0$$

The Laplace transform of this equation is

$$R_1[sI(s) - i(0)] + \frac{1}{C} I(s) = 0$$

where

$$i(0) = i_0 = \frac{E}{R_1 + R_2}$$

Hence

$$(R_1 s + \frac{1}{C})I(s) = R_1 \frac{E}{R_1 + R_2}$$

or

$$I(s) = \frac{E}{R_1 + R_2} \frac{1}{s + \frac{1}{R_1 C}}$$

The inverse Laplace transform of I(s) gives

$$i(t) = \frac{E}{R_1 + R_2} e^{-t/(R_1 C)}, \qquad t \geqslant 0$$

B-6-7. From the circuit shown to the right, we obtain

$$i_1 = \frac{e_i - e_A}{R_1}$$

$$i_2 = \frac{e_A}{R_2}$$

$$i_3 = \frac{e_A - e_o}{R_3}$$

$$i_4 = C \frac{d}{dt} (e_A - e_o)$$

$$i_5 = \frac{e_o}{R_4}$$

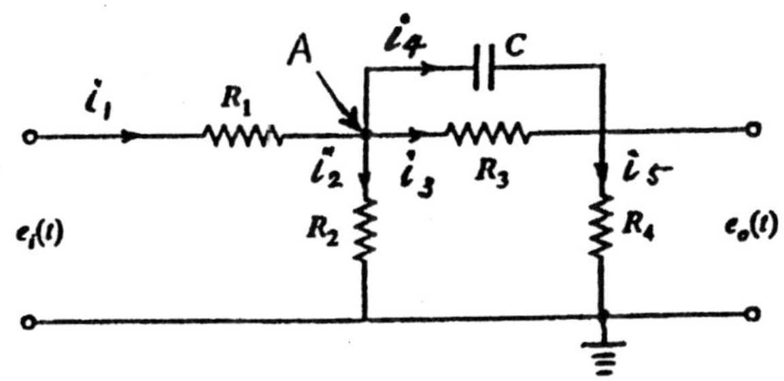

Since

$$i_1 = i_2 + i_5$$

we have

$$\frac{e_i - e_A}{R_1} = \frac{e_A}{R_2} + \frac{e_o}{R_4} \tag{1}$$

Also, since

$$i_4 + i_3 = i_5$$

we obtain

$$C \frac{d}{dt} (e_A - e_o) + \frac{e_A - e_o}{R_3} = \frac{e_o}{R_4} \tag{2}$$

Equation (1) can be written as

$$\frac{e_i}{R_1} = \left(\frac{1}{R_1} + \frac{1}{R_2}\right) e_A + \frac{e_o}{R_4}$$

Laplace transforming this equation and simplifying, we get

$$\frac{E_i(s)}{R_1} - \frac{E_o(s)}{R_4} = \left(\frac{1}{R_1} + \frac{1}{R_2}\right) E_A(s) \qquad (3)$$

Laplace transforming Equation (2), we obtain

$$Cs[E_A(s) - E_o(s)] + \frac{1}{R_3}[E_A(s) - E_o(s)] = \frac{1}{R_4} E_o(s)$$

from which we obtain

$$E_A(s) = \frac{R_3R_4Cs + R_3 + R_4}{R_4(R_3Cs + 1)} E_o(s) \qquad (4)$$

By substituting Equation (4) into Equation (3), we have

$$\frac{E_i(s)}{R_1} - \frac{E_o(s)}{R_4} = \left(\frac{R_1 + R_2}{R_1R_2}\right) \frac{R_3R_4Cs + R_3 + R_4}{R_4(R_3Cs + 1)} E_o(s)$$

from which the transfer function $E_o(s)/E_i(s)$ can be obtained as

$$\frac{E_o(s)}{E_i(s)} = \frac{R_2R_4(R_3Cs + 1)}{[R_1R_2R_3 + (R_1 + R_2)R_3R_4]Cs + R_1R_2 + (R_1 + R_2)(R_3 + R_4)}$$

B-6-8. From the circuit diagram shown, we obtain

$$\frac{E_i(s) - E_A(s)}{I_1(s)} = Ls$$

$$\frac{E_A(s)}{I_2(s)} = \frac{1}{Cs}$$

$$\frac{E_A(s) - E_o(s)}{I_3(s)} = R_1$$

$$\frac{E_o(s)}{I_4(s)} = R_2$$

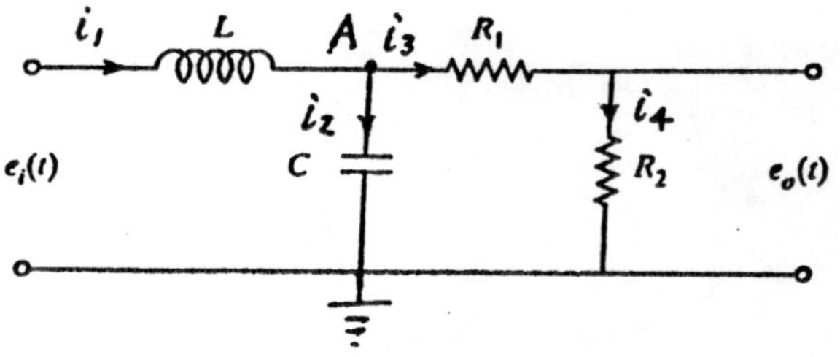

Since

$$i_1 = i_2 + i_3$$

we have

$$\frac{E_i(s) - E_A(s)}{Ls} = CsE_A(s) + \frac{E_A(s) - E_o(s)}{R_1}$$

or

$$\frac{E_i(s)}{Ls} = \left(\frac{1}{Ls} + Cs + \frac{1}{R_1}\right)E_A(s) - \frac{E_o(s)}{R_1} \tag{1}$$

Also, since

$$i_3 = i_4$$

we get

$$\frac{E_A(s) - E_o(s)}{R_1} = \frac{E_o(s)}{R_2}$$

Thus,

$$E_A(s) = \left(1 + \frac{R_1}{R_2}\right)E_o(s) \tag{2}$$

Substituting Equation (2) into Equation (1) and simplifying, we obtain

$$\frac{E_i(s)}{Ls} = \left[\left(\frac{1}{Ls} + Cs + \frac{1}{R_1}\right)R_1\left(\frac{1}{R_1} + \frac{1}{R_2}\right) - \frac{1}{R_1}\right]E_o(s)$$

$$= \left(\frac{1}{Ls} + Cs + \frac{R_1}{LR_2s} + \frac{R_1Cs}{R_2} + \frac{1}{R_2}\right)E_o(s)$$

Hence the transfer function $E_o(s)/E_i(s)$ is given by

$$\frac{E_o(s)}{E_i(s)} = \frac{R_2}{L(R_1 + R_2)Cs^2 + Ls + R_1 + R_2}$$

B-6-9. The transfer function $E_o(s)/E_i(s)$ can be given in terms of complex impedances Z_1 and Z_2 as follows:

$$\frac{E_o(s)}{E_i(s)} = \frac{Z_2}{Z_1 + Z_2}$$

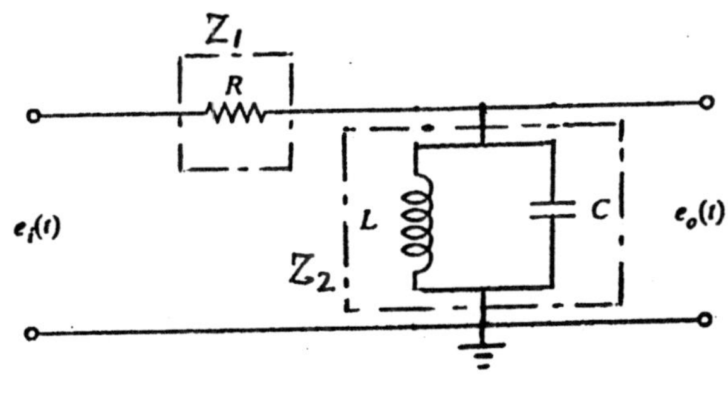

where

$$Z_1 = R$$

$$\frac{1}{Z_2} = \frac{1}{Ls} + Cs = \frac{1 + LCs^2}{Ls}$$

Hence

$$\frac{E_o(s)}{E_i(s)} = \frac{\dfrac{Ls}{LCs^2 + 1}}{R + \dfrac{Ls}{LCs^2 + 1}} = \frac{Ls}{LRCs^2 + Ls + R}$$

B-6-10. Using the impedance approach, we obtain

$$R_1 I_1(s) + Ls[I_1(s) - I_2(s)] = E_i(s) \tag{1}$$

$$R_2 I_2(s) + \frac{1}{Cs} I_2(s) + Ls[I_2(s) - I_1(s)] = 0 \tag{2}$$

$$\frac{1}{Cs} I_2(s) = E_o(s) \tag{3}$$

From Equation (2) we obtain

$$I_1(s) = \frac{LCs^2 + R_2Cs + 1}{LCs^2} \tag{4}$$

Substituting Equation (4) into Equation (1) to eliminate $I_2(s)$,

$$(R_1 + Ls) \frac{LCs^2 + R_2Cs + 1}{LCs^2} I_2(s) - LsI_2(s) = E_i(s)$$

which can be simplified to

$$[(R_1 + R_2)LCs^2 + R_1R_2Cs + Ls + R_1]I_2(s) = LCs^2E_i(s) \tag{5}$$

From Equations (3) and (5), we get

$$\frac{E_o(s)}{E_i(s)} = \frac{Ls}{(R_1 + R_2)LCs^2 + (R_1R_2C + L)s + R_1}$$

B-6-11.

$$Z_1 = R_1 + \frac{R_2}{1 + R_2Cs}, \qquad Z_2 = R_3$$

Hence

$$\frac{E_0(s)}{E_i(s)} = \frac{Z_2}{Z_1 + Z_2} = \frac{R_3}{R_1 + \dfrac{R_2}{1 + R_2Cs} + R_3}$$

$$= \frac{R_3(R_2Cs + 1)}{R_2(R_1 + R_3)Cs + R_1 + R_2 + R_3}$$

$$= \frac{R_3\left(s + \frac{1}{R_2C}\right)}{(R_1 + R_3)\left[s + \frac{R_1 + R_2 + R_3}{R_2(R_1 + R_3)C}\right]}$$

B-6-12.

$$Z_1 = Ls, \qquad \frac{1}{Z_2} = \frac{1}{R} + Cs = \frac{RCs + 1}{R}$$

Hence

$$\frac{E_o(s)}{E_i(s)} = \frac{Z_2}{Z_1 + Z_2}$$

$$= \frac{\frac{R}{RCs + 1}}{Ls + \frac{R}{RCs + 1}} = \frac{1}{LCs^2 + \frac{L}{R}s + 1}$$

B-6-13. Note that

$$Z_1 = \frac{1}{C_1 s}, \qquad Z_2 = \frac{R_2}{R_2C_2s + 1}$$

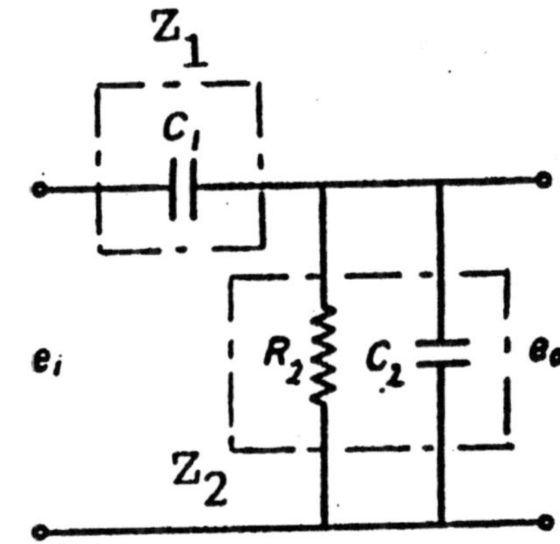

Hence

$$\frac{E_o(s)}{E_i(s)} = \frac{Z_2}{Z_1 + Z_2}$$

$$= \frac{R_2C_1s}{R_2(C_1 + C_2)s + 1}$$

which can be rewritten as

$$[(R_2C_1 + R_2C_2)s + 1]E_o(s) = R_2C_1sE_i(s)$$

which is equivalent to

$$(R_2C_1 + R_2C_2)\dot{e}_o + e_o = R_2C_1\dot{e}_i$$

or

$$\dot{e}_o + \frac{1}{R_2(C_1 + C_2)} e_o = \frac{C_1}{C_1 + C_2} \dot{e}_i$$

This equation can be written as

$$\dot{e}_o + a_1 e_o = b_0 \dot{e}_i \tag{1}$$

where

$$a_1 = \frac{1}{R_2(C_1 + C_2)}, \qquad b_0 = \frac{C_1}{C_1 + C_2}$$

Define

$$x = e_o - b_0 u, \qquad y = e_o, \qquad u = e_i$$

Then, Equation (1) becomes

$$\dot{x} + b_0\dot{u} + a_1(x + b_0 u) = b_0\dot{u}$$

or

$$\dot{x} = - a_1 x - a_1 b_0 u$$

from which the state equation can be obtained as

$$\dot{x} = - \frac{1}{R_2(C_1 + C_2)} x - \frac{C_1}{R_2(C_1 + C_2)^2} u \tag{2}$$

Noting that $y = e_o = x + b_0 u$, the output equation can be given by

$$y = x + \frac{C_1}{C_1 + C_2} u \tag{3}$$

Equations (2) and (3) give a state space representation for the given system.

B-6-14. The equations for the circuit are

$$R_1 i_1 + \frac{1}{C_1}\int i_1\, dt + L_1 \frac{di_1}{dt} = \frac{1}{C_2}\int i_2\, dt$$

$$\frac{1}{C_2}\int i_2\, dt + R_2(i_1 + i_2) = 0$$

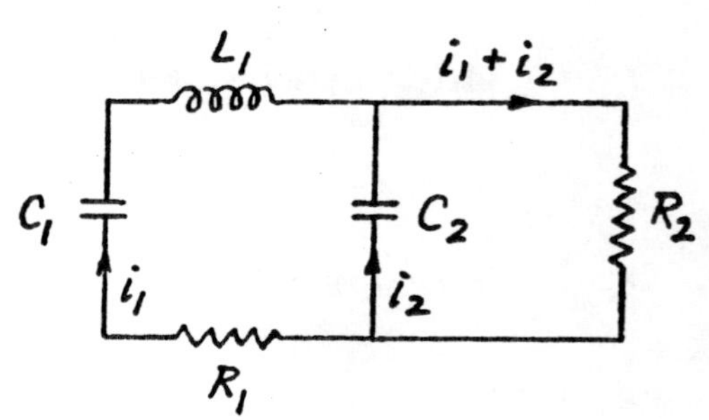

or

$$R_1\dot{q}_1 + \frac{1}{C_1} q_1 + L_1\ddot{q}_1 = \frac{1}{C_2} q_2$$

$$\frac{1}{C_2} q_2 + R_2(\dot{q}_1 + \dot{q}_2) = 0$$

Substitution of $q_1 = x_1$, $\dot{q}_1 = x_2$, $q_2 = x_3$ into the last two equations yields

$$R_1 x_2 + \frac{1}{C_1} x_1 + L_1\dot{x}_2 = \frac{1}{C_2} x_3$$

$$\frac{1}{C_2} x_3 + R_2(x_2 + \dot{x}_3) = 0$$

Hence we have

$$\dot{x}_1 = x_2$$

$$\dot{x}_2 = \frac{1}{L_1}\left(-\frac{1}{C_1} x_1 - R_1 x_2 + \frac{1}{C_2} x_3\right)$$

$$\dot{x}_3 = \frac{1}{R_2}\left(-R_2 x_2 - \frac{1}{C_2} x_3\right)$$

or

$$\begin{bmatrix} \dot{x}_1 \\ \dot{x}_2 \\ \dot{x}_3 \end{bmatrix} = \begin{bmatrix} 0 & 1 & 0 \\ -\frac{1}{L_1C_1} & -\frac{R_1}{L_1} & \frac{1}{L_1C_2} \\ 0 & -1 & -\frac{1}{R_2C_2} \end{bmatrix} \begin{bmatrix} x_1 \\ x_2 \\ x_3 \end{bmatrix}$$

B-6-15. For the mechanical system of Figure 6-81(a), the equation of motion is

$$b(\dot{x}_i - \dot{x}_o) = kx_o$$

or

$$b\dot{x}_i = kx_o + b\dot{x}_o$$

By taking the Laplace transform of this last equation, assuming zero initial conditions, we obtain

$$bsX_i(s) = (k + bs)X_o(s)$$

Hence, the transfer function between $X_o(s)$ and $X_i(s)$ is

$$\frac{X_o(s)}{X_i(s)} = \frac{bs}{bs + k} = \frac{\frac{b}{k} s}{\frac{b}{k} s + 1}$$

For the electrical system of Figure 6-81(b), we have

$$\frac{E_o(s)}{E_i(s)} = \frac{RCs}{RCs + 1}$$

Comparing the transfer functions obtained, we see that the two systems are analogous.

B-6-16. From the diagram shown here

$$Z_1 = R_1 + \frac{1}{C_1 s}$$

$$Z_2 = R_2 + \frac{1}{C_2 s}$$

Hence

$$\frac{E_o(s)}{E_i(s)} = \frac{Z_2}{Z_1 + Z_2} = \frac{R_2 + \frac{1}{C_2 s}}{R_1 + \frac{1}{C_1 s} + R_2 + \frac{1}{C_2 s}}$$

$$= \frac{R_2 C_2 s + 1}{(R_1 C_2 + R_2 C_2)s + 1 + \frac{C_2}{C_1}}$$

If we change R_1 to b_1, R_2 to b_2, C_1 to $1/k_1$, and C_2 to $1/k_2$, then we obtain

$$\frac{R_2 C_2 s + 1}{(R_1 + R_2)C_2 s + 1 + \frac{C_2}{C_1}} \longrightarrow \frac{b_2 \frac{1}{k_2} s + 1}{(b_1 + b_2) \frac{1}{k_2} s + 1 + \frac{k_1}{k_2}}$$

or

$$\frac{X_o(s)}{X_i(s)} = \frac{b_2 s + k_2}{(b_1 + b_2)s + k_2 + k_1} = \frac{b_2 s + k_2}{(b_1 s + k_1) + (b_2 s + k_2)}$$

The analogous mechanical system is shown on the next page.

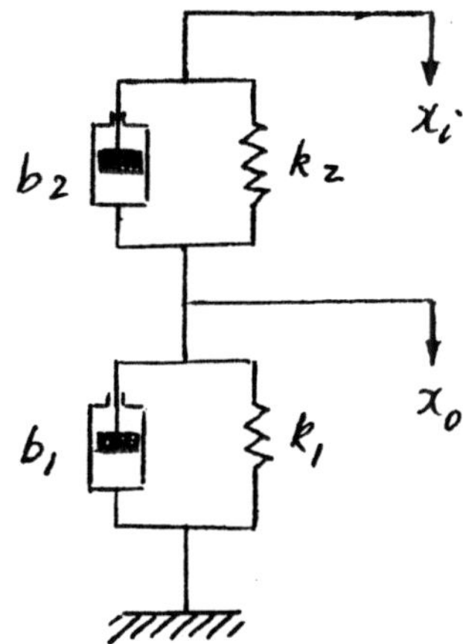

B-6-17. Define the cyclic current in the left loop as i_1 and that in the right loop as i_2. Then the equations for the circuit are

$$L_1 \frac{di_1}{dt} + \frac{1}{C_2} \int (i_1 - i_2)\, dt + R(i_1 - i_2) + \frac{1}{C_1} \int i_1 dt = 0$$

$$L_2 \frac{di_2}{dt} + \frac{1}{C_3} \int i_2\, dt + R(i_2 - i_1) + \frac{1}{C_2} \int (i_2 - i_1)\, dt = 0$$

which can be rewritten as

$$L_1 \ddot{q}_1 + \frac{1}{C_2} (q_1 - q_2) + R(\dot{q}_1 - \dot{q}_2) + \frac{1}{C_1} q_1 = 0$$

$$L_2 \ddot{q}_2 + \frac{1}{C_3} q_2 + R(\dot{q}_2 - \dot{q}_1) + \frac{1}{C_2} (q_2 - q_1) = 0$$

Using the force-voltage analogy, we can convert the last two equations as follows:

$$m_1 \ddot{x}_1 + k_2(x_1 - x_2) + b(\dot{x}_1 - \dot{x}_2) + k_1 x_1 = 0$$

$$m_2 \ddot{x}_2 + k_3 x_2 + b(\dot{x}_2 - \dot{x}_1) + k_2(x_2 - x_1) = 0$$

From these equations an analogous mechanical system can be obtained as shown to the right.

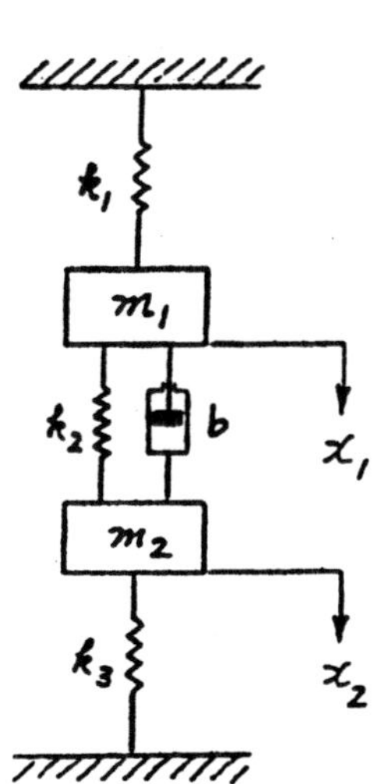

B-6-18. Define the displacement of midpoint between k_3 and b as x_3. Then the equations for the system are

$$m_1\ddot{x}_1 + k_1x_1 + k_2(x_1 - x_2) + k_3(x_1 - x_3) = p(t)$$

$$m_2\ddot{x}_2 + b(\dot{x}_2 - \dot{x}_3) + k_2(x_2 - x_1) = 0$$

$$b(\dot{x}_3 - \dot{x}_2) = k_3(x_1 - x_3)$$

Using the force-voltage analogy, the preceding equations may be converted to

$$L_1\ddot{q}_1 + \frac{1}{C_1} q_1 + \frac{1}{C_2} (q_1 - q_2) + \frac{1}{C_3}(q_1 - q_3) = e(t)$$

$$L_2\ddot{q}_2 + R(\dot{q}_2 - \dot{q}_3) + \frac{1}{C_2} (q_2 - q_1) = 0$$

$$R(\dot{q}_3 - \dot{q}_2) = \frac{1}{C_3} (q_1 - q_3)$$

The last three equations can be modified to

$$L_1\frac{di_1}{dt} + \frac{1}{C_1}\int i_1dt + \frac{1}{C_2}\int(i_1 - i_2)dt + \frac{1}{C_3}\int(i_1 - i_3)dt = e(t)$$

$$L_2\frac{di_2}{dt} + R(i_2 - i_3) + \frac{1}{C_2}\int(i_2 - i_1)dt = 0$$

$$R(i_3 - i_2) = \frac{1}{C_3}\int(i_1 - i_3)dt$$

From these three equations we can obtain the analogous electrical system as shown to the right.

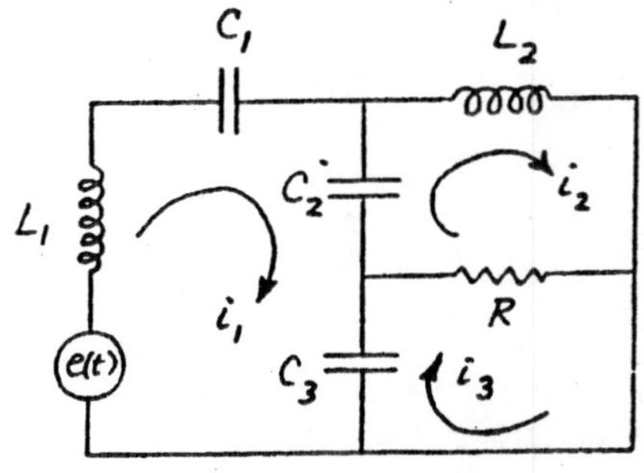

B-6-19. The equations for the system are

$$R_ai_a + L_a \frac{di_a}{dt} + e_b = e_a \quad (1)$$

$$e_b = K_b \frac{d\theta_1}{dt} \quad (2)$$

$$J_{eq} \frac{d^2\theta_1}{dt^2} + b_{eq} \frac{d\theta_1}{dt} = T = Ki_a \quad (3)$$

where J_{eq} is the equivalent moment of inertia of the system referred to the motor rotor axis and b_{eq} is the equivalent viscous friction coefficient of the system referred to the same axis.

By taking Laplace transforms of Equations (1) and (2), we obtain

$$(R_a + L_a s)I_a(s) + E_b(s) = E_a(s)$$

$$E_b(s) = K_b s\theta_1(s)$$

Elimination of $E_b(s)$ from the above two equations yields

$$(R_a + L_a s)I_a(s) + K_b s\theta_1(s) = E_a(s) \tag{4}$$

The Laplace transform of Equation (3) is

$$J_{eq}s^2\theta_1(s) + b_{eq}s\theta_1(s) = KI_a(s)$$

Hence

$$I_a(s) = \frac{J_{eq}s^2 + b_{eq}s}{K}\theta_1(s) \tag{5}$$

By substituting Equation (5) into Equation (4), we obtain

$$\left[(R_a + L_a s)\frac{J_{eq}s^2 + b_{eq}s}{K} + K_b s\right]\theta_1(s) = E_a(s)$$

or

$$\frac{\theta_1(s)}{E_a(s)} = \frac{K}{(R_a + L_a s)(J_{eq}s^2 + b_{eq}s) + KK_b s} \tag{6}$$

The numerical values of the equivalent moment of inertia J_{eq} and equivalent viscous friction coefficient b_{eq} are, respectively,

$$J_{eq} = J_r + \left(\frac{N_1}{N_2}\right)^2 J_L = 1 \times 10^{-5} + (0.1)^2 \times 4.4 \times 10^{-3}$$

$$= 5.4 \times 10^{-5} \text{ lb}_f\text{-ft-s}^2$$

$$b_{eq} = b_r + \left(\frac{N_1}{N_2}\right)^2 b_L = (0.1)^2 \times 4 \times 10^{-2}$$

$$= 4 \times 10^{-4} \text{ lb}_f\text{-ft/rad/s}$$

Substituting these numerical values into Equation (6), we get

$$\frac{\Theta_1(s)}{E_a(s)} = \frac{6 \times 10^{-5}}{0.2(5.4 \times 10^{-5}s^2 + 4 \times 10^{-4}s) + 6 \times 10^{-5} \times 5.5 \times 10^{-2}s}$$

$$= \frac{6}{1.08s^2 + 8.33s}$$

$$= \frac{0.72}{s(0.1296s + 1)}$$

Since $\theta_2/\theta_1 = N_1/N_2 = n = 0.1$, we have the transfer function $\Theta_2(s)/E_a(s)$ as follows:

$$\frac{\Theta_2(s)}{E_a(s)} = \frac{n\Theta_1(s)}{E_a(s)} = \frac{0.072}{s(0.1296s + 1)}$$

B-6-20.

$$Z_1(s) = R, \quad Z_2(s) = \frac{1}{Cs}$$

Hence

$$\frac{E_o(s)}{E_i(s)} = -\frac{Z_2(s)}{Z_1(s)} = -\frac{1}{RCs}$$

B-6-21. Define the input impedance and feedback impedance as Z_1 and Z_2, respectively. Then

$$Z_1 = \frac{1}{Cs}, \qquad Z_2 = R$$

The transfer function $E_o(s)/E_i(s)$ is given by

$$\frac{E_o(s)}{E_i(s)} = -\frac{Z_2(s)}{Z_1(s)} = -RCs$$

This circuit is a differentiator. Note that the circuit is unstable. To make the circuit stable, we add a resistor in the input path as shown in Figure 6-94 (Problem B-6-28), so that the transfer function of the circuit is of the form $-RCs/(Ts + 1)$, where T is a small time constant. (See the solution to Problem B-6-28.)

B-6-22. Define the input impedance and feedback impedance as Z_1 and Z_2, respectively. Then

$$\frac{1}{Z_1} = Cs + \frac{1}{R_1}, \qquad Z_2 = R_2$$

Hence, the transfer function $E_o(s)/E_i(s)$ is given by

$$\frac{E_o(s)}{E_i(s)} = -\frac{Z_2(s)}{Z_1(s)} = -R_2(Cs + \frac{1}{R_1}) = -\frac{R_2}{R_1}(R_1Cs + 1)$$

B-6-23. Referring to Example 6-11 or Problem A-6-21, the transfer function $E_o(s)/E_i(s)$ is given by

$$\frac{E_o(s)}{E_i(s)} = -\frac{R_2}{R_1}\frac{1}{R_2Cs + 1}$$

In the following we derive a state-space representation of the system. Since

$$(R_1R_2Cs + 1)E_o(s) = -R_2E_i(s) \qquad (1)$$

we have

$$R_1R_2C\dot{e}_o + e_o = -R_2e_i \qquad (1)$$

Define the state variable x as

$$x = e_o$$

the input variable u as

$$u = e_i$$

and the output variable y as

$$y = e_o = x$$

Then Equation (1) can be written as

$$R_1R_2C\dot{x} + x = -R_2u$$

Hence, the state equation is

$$\dot{x} = -\frac{1}{R_1R_2C}x - \frac{1}{R_1C}u$$

and the output equation is

$$y = x$$

B-6-24. Let us define the voltage at point A as e_A. Then, because negative feedback is used in the operational-amplifier circuit, the differential input voltage becomes zero. Thus we have

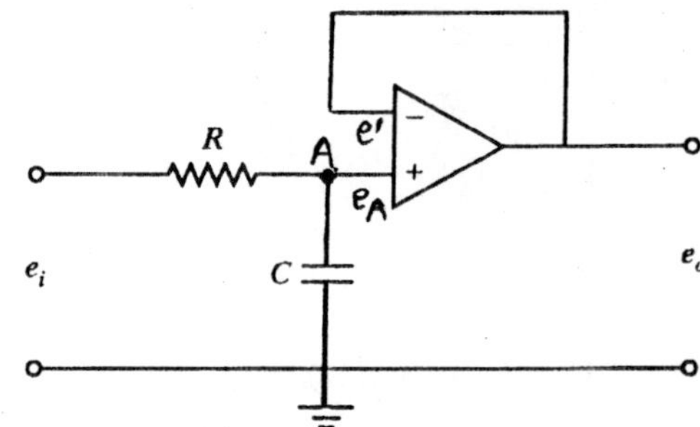

$$e_A = e'$$

Note that

$$\frac{E_A(s)}{E_i(s)} = \frac{\frac{1}{Cs}}{\frac{1}{Cs} + R} = \frac{1}{RCs + 1}$$

and

$$E'(s) = E_o(s)$$

Hence

$$E_o(s) = E_A(s) = \frac{1}{RCs + 1} E_i(s)$$

Thus the transfer function of the given operational amplifier circuit is

$$\frac{E_o(s)}{E_i(s)} = \frac{1}{RCs + 1}$$

B-6-25. Note that

$$\frac{1}{Z_1(s)} = C_1 s + \frac{1}{R_1}$$

Hence

$$Z_1(s) = \frac{R_1}{R_1 C_1 s + 1}$$

Similarly,

$$Z_2(s) = \frac{R_2}{R_2 C_2 s + 1}$$

The transfer function $E_o(s)/E_i(s)$ can be given by

$$\frac{E_o(s)}{E_i(s)} = -\frac{Z_2(s)}{Z_1(s)} = -\frac{R_2}{R_1}\frac{R_1 C_1 s + 1}{R_2 C_2 s + 1}$$

B-6-26. Define the voltage at point A as e_A. Then

$$\frac{E_A(s)}{E_i(s)} = \frac{1}{R_1Cs + 1}$$

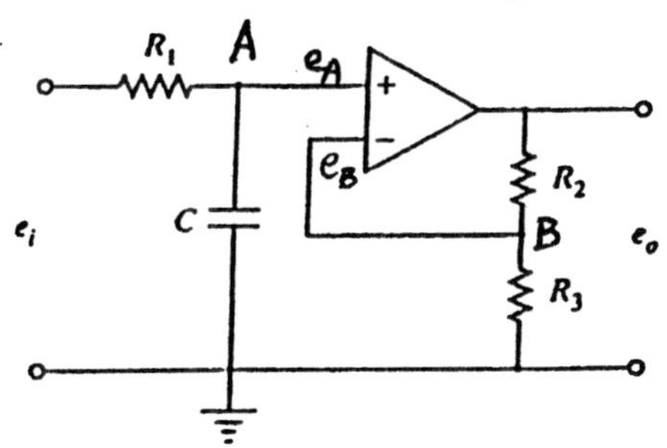

Define the voltage at point B as e_B. Then

$$E_B(s) = \frac{R_3}{R_2 + R_3} E_o(s)$$

Noting that this circuit involves negative feedback, the differential input voltage becomes zero. Thus

$$E_A(s) = E_B(s)$$

Hence

$$E_A(s) = \frac{1}{R_1Cs + 1} E_i(s) = E_B(s) = \frac{R_3}{R_2 + R_3} E_o(s)$$

from which we obtain

$$\frac{E_o(s)}{E_i(s)} = \frac{R_2 + R_3}{R_3} \frac{1}{R_1Cs + 1}$$

B-6-27. The voltage at point A is

$$e_A = \frac{1}{2}(e_i + e_o)$$

or

$$E_A(s) = \frac{1}{2}[E_i(s) + E_o(s)] \tag{1}$$

The voltage at point B is

$$E_B(s) = \frac{\frac{1}{Cs}}{R_2 + \frac{1}{Cs}} E_i(s) = \frac{1}{R_2Cs + 1} E_i(s) \tag{2}$$

Since this circuit involves negative feedback, the differential input voltage becomes zero, or

$$E_A(s) = E_B(s)$$

Thus, equating Equations (1) and (2) we obtain

$$\frac{1}{2}[E_i(s) + E_o(s)] = \frac{1}{R_2Cs + 1} E_i(s)$$

from which we get

$$\frac{E_o(s)}{E_i(s)} = -\frac{R_2Cs - 1}{R_2Cs + 1}$$

B-6-28. Define the input impedance and feedback impedance as Z_1 and Z_2, respectively. Then

$$Z_1 = R_1 + \frac{1}{Cs}$$

$$Z_2 = R_2$$

Thus,

$$\frac{E_o(s)}{E_i(s)} = -\frac{Z_2(s)}{Z_1(s)} = -\frac{R_2}{R_1 + \frac{1}{Cs}} = -\frac{R_2Cs}{R_1Cs + 1}$$

B-6-29. Define the voltage at point A as E_A. Then

$$\frac{E_A(s)}{E_i(s)} = \frac{R_1}{\frac{1}{Cs} + R_1} = \frac{R_1Cs}{R_1Cs + 1}$$

Define the voltage at point B as e_B. Then

$$E_B(s) = \frac{R_3}{R_2 + R_3} E_o(s)$$

Since this circuit involves negative feedback, the differential input voltage is equal to zero. Hence

$$E_A(s) = E_B(s)$$

Thus,

$$\frac{R_1Cs}{R_1Cs + 1} E_i(s) = E_A(s) = E_B(s) = \frac{R_3}{R_2 + R_3} E_o(s)$$

from which we obtain

$$\frac{E_o(s)}{E_i(s)} = \frac{R_2 + R_3}{R_1} \frac{R_1Cs}{R_1Cs + 1}$$

B-6-30. For the operational-amplifier circuit shown to the right, we have

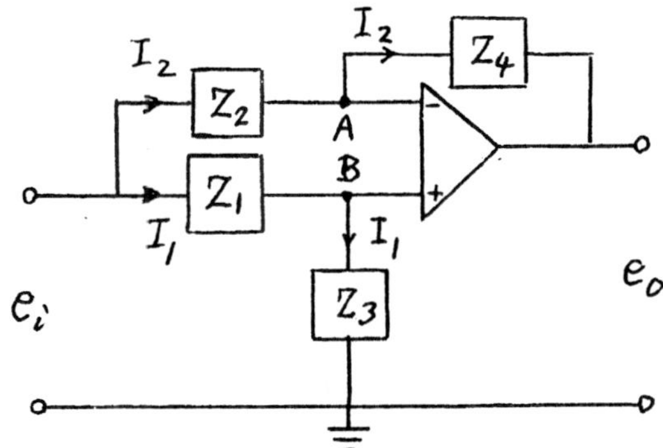

$$E_i - E_A = Z_2 I_2 \qquad (1)$$

$$E_A - E_o = Z_4 I_2 \qquad (2)$$

$$E_B = \frac{Z_3}{Z_1 + Z_3} E_i \qquad (3)$$

Since this operational-amplifier circuit involves negative feedback, the differential input voltage must be zero, or

$$E_A = E_B \qquad (4)$$

From Equations (1), (3), and (4) we have

$$E_i - E_B = E_i - \frac{Z_3}{Z_1 + Z_3} E_i = Z_2 I_2$$

Hence

$$\frac{Z_1}{Z_1 + Z_3} E_i = Z_2 I_2 \qquad (5)$$

From Equations (1) and (2), we have

$$E_i = E_o + (Z_2 + Z_4) I_2 \qquad (6)$$

By substituting Equation (6) into Equation (5), we have

$$\frac{Z_1}{Z_1 + Z_3} \left[E_o + (Z_2 + Z_4) I_2 \right] = Z_2 I_2$$

Hence

$$\frac{Z_1}{Z_1 + Z_3} E_o = \frac{Z_2 Z_3 - Z_1 Z_4}{Z_1 + Z_3} I_2 \qquad (7)$$

From Equations (5) and (7), we have

$$\frac{E_o}{E_i} = \frac{\dfrac{Z_2 Z_3 - Z_1 Z_4}{Z_1 + Z_3} I_2}{Z_2 I_2} = \frac{Z_2 Z_3 - Z_1 Z_4}{Z_1 Z_2 + Z_2 Z_3} \qquad (8)$$

For the circuit shown in Figure 6-96, we have

$$Z_1 = \frac{1}{Cs}, \quad Z_2 = R_1, \quad Z_3 = R_2, \quad Z_4 = R_1$$

Then, from Equation (8) we obtain the transfer function of the circuit to be

$$\frac{E_o(s)}{E_i(s)} = \frac{R_1R_2 - \frac{R_1}{Cs}}{\frac{1}{Cs}R_1 + R_1R_2} = \frac{R_2 - \frac{1}{Cs}}{\frac{1}{Cs} + R_2} = \frac{R_2Cs - 1}{R_2Cs + 1}$$

B-6-31. From the diagram shown to the right, we obtain

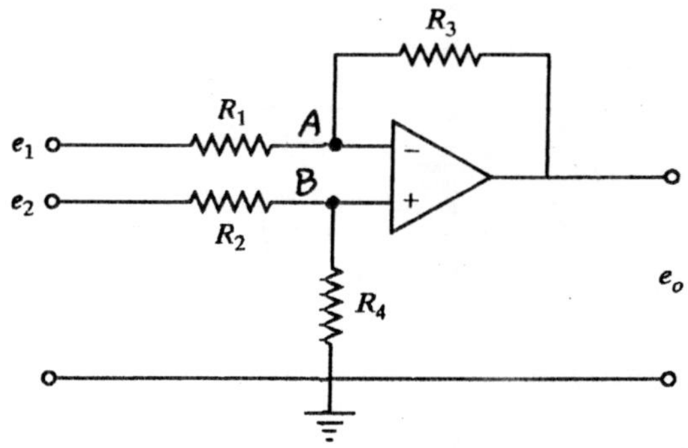

$$\frac{e_1 - e_A}{R_1} = \frac{e_A - e_o}{R_3}$$

or

$$R_3e_1 + R_1e_o = (R_3 + R_1)e_A$$

Hence

$$e_A = \frac{R_1e_o + R_3e_1}{R_1 + R_3}$$

Also

$$e_B = \frac{R_4}{R_2 + R_4} e_2$$

Since the circuit involves negative feedback, the differential input voltage is zero, or

$$e_A = e_B$$

Hence

$$\frac{R_1e_o + R_3e_1}{R_1 + R_3} = \frac{R_4}{R_2 + R_4} e_2$$

Simplifying,

$$\frac{R_1}{R_1 + R_3} e_o = \frac{R_4}{R_2 + R_4} e_2 - \frac{R_3}{R_1 + R_3} e_1$$

or

$$e_o = \frac{R_4(R_1 + R_3)}{R_1(R_2 + R_4)} e_2 - \frac{R_3}{R_1} e_1$$

This equation gives the output voltage e_o in terms of the input voltages e_1 and e_2.

Notice that if $R_1 = R_3$ and $R_2 = R_4$, then

$$e_o = e_2 - e_1$$

Thus, for this special case, the output voltage e_o is equal to the differential voltage $e_2 - e_1$.

CHAPTER 7

B-7-1. By substituting the given numerical values into

$$R_\ell = \frac{dh}{dQ} = \frac{128\nu L}{g\pi D^4}$$

we obtain

$$R_\ell = \frac{128 \times 1.004 \times 10^{-6} \times 2}{9.81 \times 3.14 \times (4 \times 10^{-3})^4} = 3.2594 \times 10^4 \text{ s/m}^2$$

B-7-2. We shall solve this problem by using two different approaches: one based on the exact method and the other based on the use of an average resistance.

(1) Solution by the exact method. For the liquid-level system we have

$$C\,dH = (Q_i - Q)\,dt$$

By substituting $C = 2\text{ m}^2$, $Q_i = 0.05\text{ m}^3/\text{s}$, and $Q = 0.02\sqrt{H}$ into this last equation, we obtain

$$2\,dH = (0.05 - 0.02\sqrt{H})\,dt$$

or

$$dt = \frac{2}{0.05 - 0.02\sqrt{H}}\,dH$$

Let $\sqrt{H} = x$. Then $H = x^2$ and $dH = 2x\,dx$. So we have

$$dt = \frac{200}{5 - 2x}\,2x\,dx = 200\left(\frac{-5 + 2x + 5}{5 - 2x}\right)dx$$

Assume that at $t = t_1$ the level reaches 2.5 m. Then, t_1 is obtained as

$$t_1 = \int_0^{t_1} dt = \int_1^{\sqrt{2.5}} 200\left(-1 + \frac{5}{5 - 2x}\right)dx$$

$$= -200\,x\Big|_1^{\sqrt{2.5}} + 1000\int_1^{\sqrt{2.5}}\frac{dx}{5 - 2x}$$

$$= -200(\sqrt{2.5} - 1) + 1000\left(-\frac{1}{2}\right)\ln(5 - 2x)\Big|_1^{\sqrt{2.5}}$$

$$= -200 \times 0.581 - 500(\ln 1.838 - \ln 3)$$

$$= -200 \times 0.581 - 500(0.6087 - 1.0986) = 128.8\text{ s}$$

(2) Solution by use of an average resistance. Since $Q = 0.02\sqrt{H}$, the average resistance R is obtained from

$$R = \frac{dH}{dQ} = \frac{2.5 - 1.0}{0.03162 - 0.02} = 129$$

Define $h = H - 1$. Then $q_i = Q_i - 0.02$ and $q_o = h/R$. For the liquid-level system,

$$C\,dh = (q_i - q_o)dt$$

or

$$C\frac{dh}{dt} = q_i - q_o = q_i - \frac{h}{R}$$

which can be rewritten as

$$CR\frac{dh}{dt} + h = q_i R$$

Substituting $C = 2\ m^2$, $R = 129\ s/m^2$, and $q_i = 0.05 - 0.02 = 0.03\ m^3/s$ into this last equation yields

$$258\frac{dh}{dt} + h = 3.87, \qquad h(0) = 0$$

Taking Laplace transforms of both sides of this last equation, we obtain

$$258[sH(s) - h(0)] + H(s) = \frac{3.87}{s}$$

or

$$(258s + 1)H(s) = \frac{3.87}{s}$$

Solving this equation for H(s),

$$H(s) = \frac{3.87}{s(258s + 1)}$$

$$= 3.87\left(\frac{1}{s} - \frac{258}{258s + 1}\right)$$

The inverse Laplace transform of this last equation gives

$$h(t) = 3.87(1 - e^{-\frac{1}{258}t})$$

Assume that at $t = t_1$, $h(t_1) = 1.5$. The value of t_1 can be determined from

$$1.5 = 3.87(1 - e^{-\frac{1}{258}t_1})$$

Rewriting,

$$e^{-\frac{1}{258}t_1} = 0.6124$$

or

$$\frac{t_1}{258} = 0.4904$$

So we have

$$t_1 = 0.4904 \times 258 = 126.5 \text{ s}$$

This solution has been obtained by use of an average resistance.

B-7-3. The equations for the liquid-level system are

$$C_1 dh_1 = (\bar{Q} + q - \bar{Q} - q_1)dt$$

$$C_2 dh_2 = (\bar{Q} + q_1 - \bar{Q} - q_2)dt$$

Since $R_1 = h_1/q_1$ and $R_2 = h_2/q_2$, the system equations can be rewritten as

$$C_1 \frac{dh_1}{dt} = q - q_1 = q - \frac{h_1}{R_1} \qquad (1)$$

$$C_2 \frac{dh_2}{dt} = q_1 - q_2 = \frac{h_1}{R_1} - \frac{h_2}{R_2} \qquad (2)$$

From Equations (1) and (2) we obtain

$$C_1 \frac{dh_1}{dt} + C_2 \frac{dh_2}{dt} = q - \frac{h_2}{R_2} \qquad (3)$$

By differentiating Equation (2) with respect to t, we get

$$C_2 \frac{d^2h_2}{dt^2} + \frac{1}{R_2}\frac{dh_2}{dt} = \frac{1}{R_1}\frac{dh_1}{dt} \qquad (4)$$

By eliminating dh_1/dt from Equations (3) and (4), we obtain

$$R_1C_1R_2C_2 \frac{d^2h_2}{dt^2} + (R_1C_1 + R_2C_2)\frac{dh_2}{dt} + h_2 = R_2q$$

Substitution of $h_2 = R_2q_2$ into this last equation yields

$$R_1C_1R_2C_2 \frac{d^2q_2}{dt^2} + (R_1C_1 + R_2C_2)\frac{dq_2}{dt} + q_2 = q$$

Hence the transfer function of the system when q is the input and q_2 is the output is given by

$$\frac{Q_2(s)}{Q(s)} = \frac{1}{(R_1C_1s + 1)(R_2C_2s + 1)}$$

B-7-4. The equations for the system are

$$C_1dh_1 = q_1dt$$

$$C_2dh_2 = (q_i - q_1 - q_2)dt$$

$$C_3dh_3 = (q_2 - q_0)dt$$

where

$$q_1 = \frac{h_2 - h_1}{R_1}$$

$$q_2 = \frac{h_2 - h_3}{R_2}$$

$$q_0 = \frac{h_3}{R_3}$$

Thus, we have

$$C_1\frac{dh_1}{dt} = \frac{h_2 - h_1}{R_1} \qquad (1)$$

$$C_2\frac{dh_2}{dt} = q_i - \frac{h_2 - h_1}{R_1} - \frac{h_2 - h_3}{R_2} \qquad (2)$$

$$C_3\frac{dh_3}{dt} = \frac{h_2 - h_3}{R_2} - \frac{h_3}{R_3} \qquad (3)$$

From Equation (1) we obtain

$$C_1sH_1(s) = \frac{1}{R_1}[H_2(s) - H_1(s)]$$

or

$$H_1(s) = \frac{1}{R_1C_1s + 1}H_2(s) \qquad (4)$$

From Equation (3) we get

$$C_3R_3sH_3(s) + H_3(s) = \frac{R_3}{R_2} H_2(s) - \frac{R_3}{R_2} H_3(s)$$

or

$$H_2(s) = \frac{R_2}{R_3} (R_3C_3s + 1 + \frac{R_3}{R_2})H_3(s) \tag{5}$$

By adding Equations (1), (2), and (3), and taking the Laplace transform of the resulting equation, we obtain

$$C_1sH_1(s) + C_2sH_2(s) + C_3sH_3(s) = Q_i(s) - \frac{1}{R_3} H_3(s) \tag{6}$$

By substituting Equations (4) and (5) into Equation (6), we get

$$\left[\left(\frac{C_1s}{R_1C_1s + 1} + C_2s\right)\left(\frac{R_2}{R_3}\right)(R_3C_3s + 1 + \frac{R_3}{R_2}) + (C_3s + \frac{1}{R_3})\right]H_3(s) = Q_i(s)$$

Since $H_3(s) = R_3Q_o(s)$, this last equation can be written as

$$\left[\frac{(C_1 + C_2)s + R_1C_1C_2s^2}{R_1C_1s + 1}(R_3R_2C_3s + R_2 + R_3) + (R_3C_3s + 1)\right] Q_o(s) = Q_i(s)$$

from which we obtain

$$\frac{Q_o(s)}{Q_i(s)} = \frac{R_1C_1s + 1}{[(C_1 + C_2)s + R_1C_1C_2s^2](R_3R_2C_3s + R_2 + R_3) + (R_3C_3s + 1)(R_1C_1s + 1)}$$

This is the transfer function relating $Q_o(s)$ and $Q_i(s)$.

B-7-5. For this system

$$C dH = -Q\,dt, \qquad H = 3r, \qquad C = r^2\pi = \left(\frac{H}{3}\right)^2 \pi$$

Hence

$$\left(\frac{H}{3}\right)^2 \pi\, dH = -0.005\sqrt{H}\, dt$$

or

$$H^{1.5}\, dH = -0.005 \frac{9}{\pi} dt$$

Assume that the head moves down from H = 2m to x for the 60 second period. Then

$$\int_2^x H^{1.5}\, dH = -0.005 \frac{9}{\pi}\int_0^{60} dt$$

or

$$\frac{2}{5}(x^{2.5} - 2^{2.5}) = -0.01432(60 - 0)$$

which can be rewritten as

$$x^{2.5} - (1.414213)^5 = -2.1480$$

or

$$x^{2.5} = 5.6569 - 2.1480 = 3.5089$$

Taking logarithm of both sides of this last equation, we obtain

$$2.5 \log_{10} x = \log_{10} 3.5089$$

or

$$x = 1.652 \text{ m}$$

B-7-6. From Figure 7-30 we obtain

$$C_1 \frac{dh_1}{dt} = q - q_1$$

$$C_2 \frac{dh_2}{dt} = q_1 - q_2$$

$$q_1 = \frac{h_1}{R_1}$$

$$q_2 = \frac{h_2}{R_2}$$

Using the electrical-liquid-level analogy given below, equations for an analogous electrical system can be obtained.

Electrical systems	Liquid-level systems
e (voltage)	q (flow rate)
q (charge)	h (head)
i (current)	dh/dt
C (capacitance)	R (resistance)
R (resistance)	C (capacitance)

Analogous equations for the electrical system are

$$R_1 i_1 = e - e_1 \tag{1}$$

$$R_2 i_2 = e_1 - e_2 \tag{2}$$

$$e_1 = \frac{\int i_1 \, dt}{C_1} \tag{3}$$

$$e_2 = \frac{\int i_2 \, dt}{C_2} \tag{4}$$

Based on Equations (1) through (4), we obtain the analogous electrical system shown below.

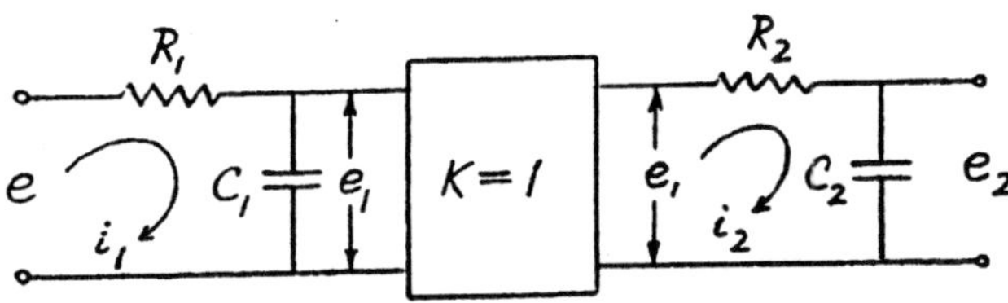

B-7-7. The equations for the liquid-level system of Figure 7-21 are

$$C_1 \frac{dh_1}{dt} = q - q_1$$

$$C_2 \frac{dh_2}{dt} = q_1 - q_2$$

$$q_1 = \frac{h_1 - h_2}{R_1}$$

$$q_2 = \frac{h_2}{R_2}$$

Using the table of electrical-liquid-level analogy shown in the solution of Problem B-7-6, we can obtain an analogous electrical system. The analogous electrical equations are

$$R_1 i_1 = e - e_1 \tag{1}$$

$$R_2 i_2 = e_1 - e_2 \tag{2}$$

$$e_1 = \frac{\int (i_1 - i_2) \, dt}{C_1} \tag{3}$$

$$e_2 = \frac{\int i_2 \, dt}{C_2} \tag{4}$$

Based on Equations (1) through (4), we obtain the analogous electrical system shown on next page.

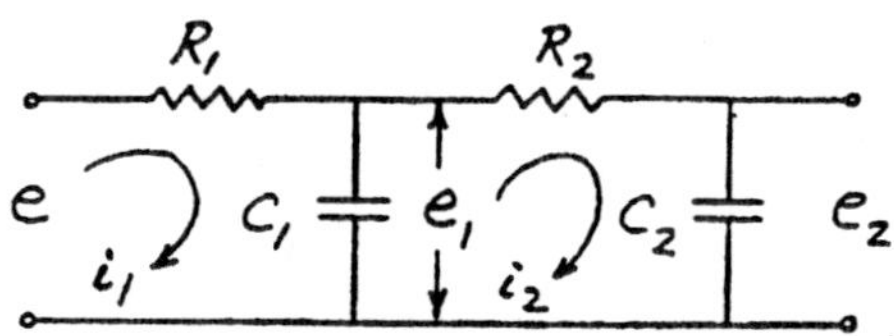

B-7-8.

$$pV = mR_{air}T$$

In this problem

$$p = 7 \text{ X } 10^5 + 1.0133 \times 10^5 = 8.0133 \times 10^5 \text{ N/m}^2 \text{ abs}$$

$$T = 273 + 20 = 293 \text{ K}$$

The mass m of the air in the tank is

$$m = \frac{pV}{R_{air}T} = \frac{8.0133 \times 10^5 \times 10}{287 \times 293} = 95.29 \text{ kg}$$

If the temperature of compressed air is raised to 40°C, then T = 273 + 40 = 313 K and the pressure p becomes

$$p = \frac{mR_{air}T}{V} = \frac{95.29 \times 287 \times 313}{10} = 8.560 \times 10^5 \text{ N/m}^2 \text{ abs}$$

$$= 7.547 \times 10^5 \text{ N/m}^2 \text{ gage} = 7.695 \text{ kg}_f/\text{cm}^2 \text{ gage}$$

$$= 109.4 \text{ lb}_f/\text{in.}^2 \text{ gage}$$

B-7-9. Note that

$$C\, dp_o = q\, dt$$

where q is the flow rate through the valve and is given by

$$q = \frac{p_i - p_o}{R}$$

Hence

$$C \frac{dp_o}{dt} = \frac{p_i - p_o}{R}$$

from which we obtain

$$\frac{P_o(s)}{P_i(s)} = \frac{1}{RCs + 1}$$

For the bellows and spring, we have the following equation:

$$Ap_o = kx$$

The transfer function $X(s)/P_i(s)$ is then given by

$$\frac{X(s)}{P_i(s)} = \frac{X(s)}{P_o(s)} \frac{P_o(s)}{P_i(s)} = \frac{A}{k} \frac{1}{RCs + 1}$$

B-7-10. Note that for $t \leqslant 0$,

$$p_1 = 0.5 \times 10^5 \text{ N/m}^2 \text{ gage} = 1.5133 \times 10^5 \text{ N/m}^2 \text{ abs}$$

$$p_2 = 0 \text{ N/m}^2 \text{ gage} = 1.0133 \times 10^5 \text{ N/m}^2 \text{ abs}$$

For $t > 0$, p_2 increases. Since p_1 stays constant for all t, we have

$$p_2(t \geqslant 0) > 0.528\, p_1$$

Note that for $p_2 > 0.528\, p_1$, the speed of air flow is subsonic. So the flow throughout the system is subsonic. The flow rate through the inlet valve is

$$q_1 = K_1\sqrt{p_1 - p_2}$$

The flow rate through the outlet valve is

$$q_2 = K_2\sqrt{p_2 - p_3}$$

Since both valves have identical flow characteristics, we have $K_1 = K_2 = K$. The equation for the system is

$$C\, dp_2 = (q_1 - q_2)dt$$

or

$$C\frac{dp_2}{dt} = K\sqrt{p_1 - p_2} - K\sqrt{p_2 - p_3}$$

At steady state, we have $dp_2/dt = 0$ and this last equation becomes

$$K\sqrt{p_1 - p_2} = K\sqrt{p_2 - p_3}$$

or

$$p_1 - p_2 = p_2 - p_3$$

Hence

$$p_2 = \frac{p_1 + p_3}{2} = \frac{1.5133 \times 10^5 + 1.0133 \times 10^5}{2}$$

$$= 1.2633 \times 10^5 \text{ N/m}^2 \text{ abs} = 0.25 \times 10^5 \text{ N/m}^2 \text{ gage}$$

B-7-11. For the toggle joint shown in Figure 7-35, we have

$$\frac{R}{\frac{1}{2}F} = \frac{\ell_2}{\ell_1}$$

Hence

$$F = 2\frac{\ell_1}{\ell_2}R$$

B-7-12. Assume that the minimum area of the piston is A m^2. Then the minimum force needed to move the load mass is

$$F = A(p_1 - p_2) = \mu mg$$

Hence

$$A = \frac{\mu mg}{p_1 - p_2} = \frac{0.3 \times 1000 \times 9.807}{5 \times 10^5}$$

$$= 588.42 \times 10^{-5} = 0.0058842 \text{ m}^2$$

Thus the minimum area of the piston is 58.84 cm^2.

B-7-13. If we assume that the mass of the power piston is negligible compared with mass m, then the equation for the system is

$$A(p_1 - p_2) - mg \sin\alpha - \mu mg \cos\alpha = m\ddot{x}$$

The load can be pushed upward if

$$A(p_1 - p_2) - mg \sin\alpha - \mu mg \cos\alpha \geqslant 0$$

Noting that $\theta = \tan^{-1}\mu$, we have $\tan\theta = \mu$. Hence

$$A(p_1 - p_2) \geqslant mg(\sin\alpha + \mu\cos\alpha) = mg(\sin\alpha + \tan\theta\cos\alpha)$$

$$= mg\frac{\cos\theta\sin\alpha + \sin\theta\cos\alpha}{\cos\theta}$$

$$= mg\frac{\sin(\theta + \alpha)}{\cos\theta}$$

Thus

$$A \geqslant \frac{mg \sin (\theta + \alpha)}{(p_1 - p_2) \cos \theta}$$

B-7-14. Define the radius and angle of rotation of the pinion as r and θ, respectively. Then, relative displacement between rack C and pinion B is $r\theta$.

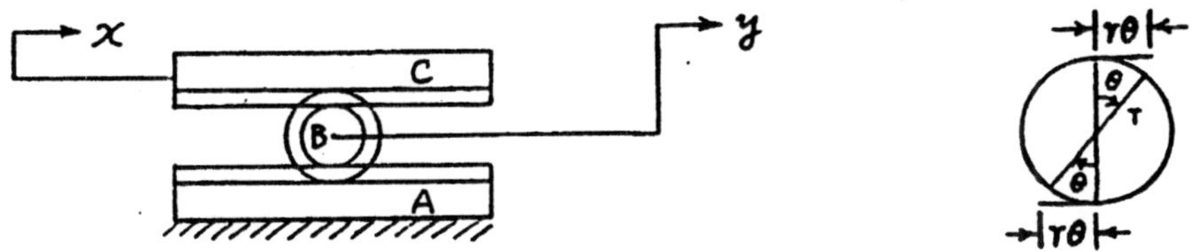

Relative displacement between rack A and rack C is $2r\theta$ and this must equal displacement x. Therefore, we have

$$2r\theta = x$$

Since $x = r\theta + y$, we obtain

$$y = \frac{x}{2}$$

B-7-15.

$$Q = 0.1\sqrt{H} = f(H)$$

$$= f(\bar{H}) + \left.\frac{df}{dH}\right|_{H = \bar{H}} (H - \bar{H}) + \frac{1}{2!} \left.\frac{d^2f}{dH^2}\right|_{H = \bar{H}} (H - \bar{H})^2 + \cdots$$

Neglecting the higher-order terms, a linearized equation for the system can be written as

$$Q - f(\bar{H}) = a(H - \bar{H})$$

where

$$f(\bar{H}) = f(4) = 0.2$$

$$a = \left.\frac{df}{dH}\right|_{H = \bar{H} = 4} = 0.1 \frac{1}{2\sqrt{\bar{H}}} = 0.025$$

Thus, a linearized equation becomes

$$Q - 0.2 = 0.025(H - 4)$$

B-7-16.

$$z = 5x^2 = f(x)$$

$$= f(\bar{x}) + \frac{df}{dx}(x - \bar{x}) + \frac{1}{2!}\frac{d^2f}{dx^2}(x - \bar{x})^2 + \cdots$$

A linearized equation for the system is

$$z - \bar{z} = a(x - \bar{x})$$

where $\bar{x} = 2$, $\bar{z} = 20$, and

$$a = \left.\frac{df}{dx}\right|_{x = 2,\ z = 20} = 10\,x \Big|_{x = 2,\ z = 20} = 20$$

Thus, a linearized equation becomes

$$z - 20 = 20(x - 2)$$

or

$$z - 20x = -20$$

B-7-17.

$$z = x^2 + 2xy + 5y^2 = f(x, y)$$

A linearized mathematical model is

$$z - \bar{z} = \frac{\partial f}{\partial x}(x - \bar{x}) + \frac{\partial f}{\partial y}(y - \bar{y})$$

where $\bar{x} = 11$, $\bar{y} = 5$, $\bar{z} = 356$, and

$$\frac{\partial f}{\partial x} = 2x + 2y \Big|_{x = 11,\ y = 5} = 22 + 10 = 32$$

$$\frac{\partial f}{\partial y} = 2x + 10y \Big|_{x = 11,\ y = 5} = 22 + 50 = 72$$

Thus, the linearized equation is

$$z - 356 = 32(x - 11) + 72(y - 5)$$

or

$$32x + 72y - z = 356$$

B-7-18.

$$P_2 = \frac{A_2}{A_1} P_1, \qquad x_2 = \frac{A_1}{A_2} x_1$$

B-7-19. Define displacements e, x, and y as shown in the figure below.

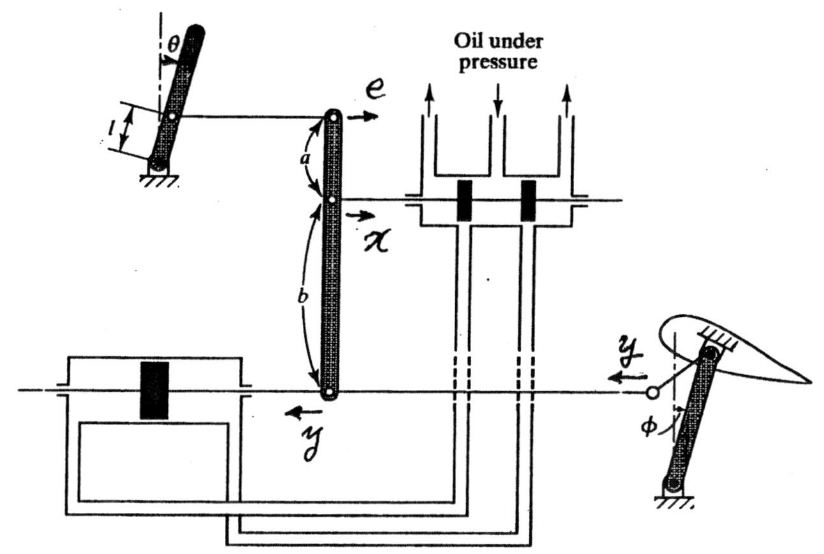

From the figure, we can construct a block diagram as shown below.

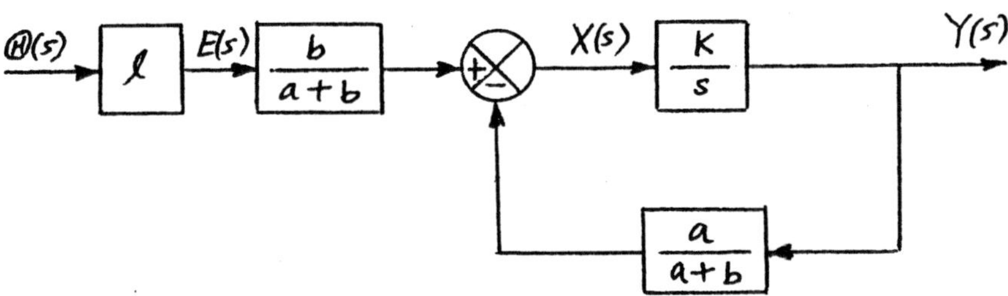

From the block diagram we obtain the transfer function Y(s)/Θ(s) as follows:

$$\frac{Y(s)}{\Theta(s)} = \frac{\ell \dfrac{b}{a+b} \dfrac{K}{s}}{1 + \dfrac{K}{s} \dfrac{a}{a+b}} \doteqdot \frac{\ell b}{a+b} \frac{a+b}{a} = \ell \frac{b}{a}$$

We see that the piston displacement y is proportional to the deflection angle θ of the control lever. Also, from the system diagram we see that for each value of y, there is a corresponding value of angle ϕ. Therefore, for each angle θ of the control lever, there is a corresponding steady-state elevator angle ϕ.

B-7-20. The heat balance equations for the system are

$$C_1 d\theta_1 = (u - q_1)dt \tag{1}$$

$$C_2 d\theta_2 = (q_1 - q_2)dt \tag{2}$$

Noting that

$$q_1 = Gc\theta_1, \qquad q_2 = Gc\theta_2$$

Equations (1) and (2) can be modified to

$$C_1 \frac{d\theta_1}{dt} = u - Gc\theta_1 \tag{3}$$

$$C_2 \frac{d\theta_2}{dt} = Gc\theta_1 - Gc\theta_2 \tag{4}$$

from which we get

$$C_1 s\Theta_1(s) = U(s) - Gc\Theta_1(s)$$

$$C_2 s\Theta_2(s) = Gc\Theta_1(s) - Gc\Theta_2(s)$$

By eliminating $\Theta_1(s)$ from the preceding two equations, we obtain

$$(C_2 s + Gc)\Theta_2(s) = \frac{Gc}{C_1 s + Gc} U(s)$$

or

$$(C_1 s + Gc)(C_2 s + Gc)\Theta_2(s) = GcU(s)$$

Thus the transfer function $\Theta_2(s)/U(s)$ can be given by

$$\frac{\Theta_2(s)}{U(s)} = \frac{Gc}{(C_1 s + Gc)(C_2 s + Gc)}$$

CHAPTER 8

B-8-1. Define the current in the circuit as i(t), where $t \geqslant 0$. The equation for the circuit for $t \geqslant 0$ is

$$(R_1 + R_2)i + \frac{1}{C_2}\int i\,dt = E$$

Since the capacitor is not charged for $t < 0$, the Laplace transform of this equation becomes

$$(R_1 + R_2)I(s) + \frac{1}{C_2 s} I(s) = \frac{E}{s}$$

Hence

$$I(s) = \frac{E}{R_1 + R_2 + \frac{1}{C_2 s}} \frac{1}{s} = \frac{EC_2}{(R_1 + R_2)C_2 s + 1}$$

Since

$$E_o(s) = (R_2 + \frac{1}{C_2 s})I(s)$$

we obtain

$$E_o(s) = \frac{R_2 C_2 s + 1}{C_2 s} \frac{EC_2}{(R_1 + R_2)C_2 s + 1}$$

$$= E\left[\frac{1}{s} - \frac{C_2 R_1}{(R_1 + R_2)C_2 s + 1}\right]$$

$$= E\left[\frac{1}{s} - \frac{R_1}{R_1 + R_2} \frac{1}{s + \frac{1}{(R_1 + R_2)C_2}}\right]$$

The inverse Laplace transform of $E_o(s)$ gives

$$e_o(t) = E\left\{1 - \frac{R_1}{R_1 + R_2} e^{-t/[(R_1 + R_2)C_2]}\right\}$$

B-8-2. The equation for the circuit is

$$L\frac{di}{dt} + Ri + \frac{1}{C}\int i\,dt = E$$

or

$$L \frac{d^2q}{dt^2} + R \frac{dq}{dt} + \frac{1}{C} q = E$$

Since $q(0) = 0$ and $i(0) = \dot{q}(0) = 0$, the Laplace transform of this last equation gives

$$Ls^2Q(s) + RsQ(s) + \frac{1}{C} Q(s) = \frac{E}{s}$$

or

$$Q(s) = \frac{E}{s(Ls^2 + Rs + \frac{1}{C})}$$

Since the current $i(t)$ is $dq(t)/dt$, we have

$$I(s) = sQ(s) = \frac{E}{Ls^2 + Rs + \frac{1}{C}}$$

The current $i(t)$ will be oscillatory if the two roots of the characteristic equation

$$s^2 + \frac{R}{L} s + \frac{1}{LC} = 0$$

are complex conjugate. If two roots are real, then the current is not oscillatory.

Case 1 (Two roots of the characteristic equation are complex conjugate):

For this case, define

$$\omega_n = \sqrt{\frac{1}{LC}}, \qquad \zeta = \frac{R\sqrt{C}}{2\sqrt{L}}$$

Then

$$I(s) = EC \frac{\omega_n^2}{s^2 + 2\zeta\omega_n s + \omega_n^2}$$

The inverse Laplace transform of $I(s)$ gives

$$i(t) = EC \frac{\omega_n}{\sqrt{1-\zeta^2}} e^{-\zeta\omega_n t} \sin(\omega_n\sqrt{1-\zeta^2}\, t)$$

$$= \frac{E\sqrt{C}}{\sqrt{L}} \frac{1}{\sqrt{1-\zeta^2}} e^{-\zeta\omega_n t} \sin(\omega_n \sqrt{1-\zeta^2}\, t) \quad (t \geq 0)$$

The current $i(t)$ approaches zero as t approaches infinity.

Case 2 (Two roots of the characteristic equation are real):

For this case define

$$s^2 + \frac{R}{L} s + \frac{1}{LC} = (s + a)(s + b)$$

Then

$$I(s) = \frac{E}{L} \frac{1}{(s + a)(s + b)}$$

The inverse Laplace transform of $I(s)$ gives

$$i(t) = \frac{E}{L} \frac{1}{b - a} (e^{-at} - e^{-bt})$$

Notice that

$$a = \frac{R}{2L} + \sqrt{\left(\frac{R}{2L}\right)^2 - \frac{1}{LC}}, \qquad b = \frac{R}{2L} - \sqrt{\left(\frac{R}{2L}\right)^2 - \frac{1}{LC}}$$

Hence

$$\frac{E}{L} \frac{1}{b - a} = - \frac{E}{2\sqrt{\left(\frac{R}{2}\right)^2 - \frac{L}{C}}}$$

The current $i(t)$ can thus be given by

$$i(t) = \frac{-E}{2\sqrt{\left(\frac{R}{2}\right)^2 - \frac{L}{C}}} \left\{ \exp\left[-\left(\frac{R}{2L} + \sqrt{\left(\frac{R}{2L}\right)^2 - \frac{1}{LC}}\right)t\right] - \exp\left[-\left(\frac{R}{2L} - \sqrt{\left(\frac{R}{2L}\right)^2 - \frac{1}{LC}}\right)t\right] \right\} \qquad (t \geqslant 0)$$

B-8-3. Referring to the circuit diagram shown to the right, we have

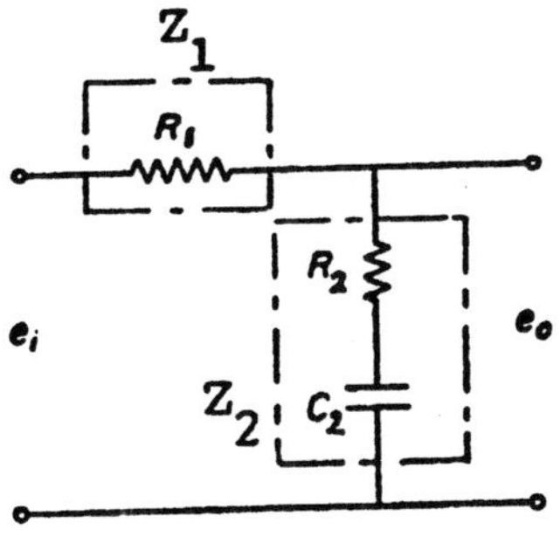

$$Z_1 = R_1, \qquad Z_2 = R_2 + \frac{1}{C_2 s}$$

Hence

$$\frac{E_o(s)}{E_i(s)} = \frac{Z_2}{Z_1 + Z_2} = \frac{R_2C_2s + 1}{(R_1 + R_2)C_2s + 1}$$

Next, we shall find the response $e_o(t)$ when the input $e_i(t)$ is the step function of magnitude E_i. Since

$$E_o(s) = \frac{R_2C_2s + 1}{(R_1 + R_2)C_2s + 1} \frac{E_i}{s}$$

$$= E_i \left[\frac{1}{s} - \frac{R_1C_2}{(R_1 + R_2)C_2s + 1} \right]$$

the inverse Laplace transform of $E_o(s)$ is

$$e_o(t) = E_i \left\{ 1 - \frac{R_1}{R_1 + R_2} e^{-t/[(R_1 + R_2)C_2]} \right\}$$

which gives the response to the step input of magnitude E_i.

B-8-4. The system equations are

$$k_1(x_i - y) = b_1(\dot{y} - \dot{x}_o)$$

$$b_1(\dot{y} - \dot{x}_o) = k_2x_o$$

which can be rewritten as

$$b_1\dot{y} + k_1y = k_1x_i + b_1\dot{x}_o$$

$$b_1\dot{x}_o + k_2x_o = b_1\dot{y}$$

Noting that $x_o(0-) = 0$ and $y(0-) = 0$, by taking the $\mathcal{L}_-$ transform of these two equations we obtain

$$(b_1s + k_1)Y(s) = k_1X_i(s) + b_1sX_o(s)$$

$$(b_1s + k_2)X_o(s) = b_1sY(s)$$

By eliminating Y(s) from the preceding two equations, we get

$$(b_1s + k_1) \frac{b_1s + k_2}{b_1s} X_o(s) = k_1X_i(s) + b_1sX_o(s)$$

Simplifying,

$$[(k_1 + k_2)b_1s + k_1k_2]\, X_o(s) = k_1b_1sX_i(s)$$

or

$$\frac{X_o(s)}{X_i(s)} = \frac{k_1b_1s}{(k_1 + k_2)b_1s + k_1k_2} = \frac{\frac{b_1}{k_2}s}{\left(\frac{1}{k_1} + \frac{1}{k_2}\right)b_1s + 1}$$

The response of the system to $x_i(t) = X_i 1(t)$ can be obtained by taking the inverse Laplace transform of

$$X_o(s) = \frac{(b_1/k_2)s}{\left(\frac{1}{k_1} + \frac{1}{k_2}\right)b_1 s + 1} \frac{X_i}{s}$$

$$= \frac{k_1 X_i}{k_1 + k_2} \frac{1}{s + \frac{1}{\left(\frac{1}{k_1} + \frac{1}{k_2}\right)b_1}}$$

as follows:

$$x_o(t) = \frac{k_1 X_i}{k_1 + k_2} \exp\left\{-t/[(\frac{1}{k_1} + \frac{1}{k_2})b_1]\right\} \qquad t \geqslant 0$$

B-8-5. The equations of motion for the system are

$$k_1(x_i - x_o) = b_2(\dot{x}_o - \dot{y})$$

$$b_2(\dot{x}_o - \dot{y}) = k_2 y$$

Rewriting these equations,

$$b_2\dot{x}_o + k_1 x_o = k_1 x_i + b_2\dot{y}$$

$$b_2\dot{y} + k_2 y = b_2\dot{x}_o$$

Noting that $x(0-) = 0$ and also $y(0-) = 0$, $\mathcal{L}_-$ transforms of the last two equations give

$$(b_2 s + k_1)X_o(s) = k_1 X_i(s) + b_2 sY(s)$$

$$(b_2 s + k_2)Y(s) = b_2 sX_o(s)$$

Eliminating Y(s) from the last two equations, we obtain

$$(b_2 s + k_1)X_o(s) = k_1 X_i(s) + b_2 s \frac{b_2 sX_o(s)}{b_2 s + k_2}$$

which can be simplified as

$$[(k_1 + k_2)b_2 s + k_1 k_2]X_o(s) = k_1(b_2 s + k_2)X_i(s)$$

Hence

$$\frac{X_o(s)}{X_i(s)} = \frac{k_1(b_2 s + k_2)}{(k_1 + k_2)b_2 s + k_1 k_2}$$

Since the input $x_i(t)$ is given as

$$x_i(t) = X_i \qquad 0 < t < t_1$$
$$= 0 \qquad \text{elsewhere}$$

we have

$$X_i(s) = \frac{X_i}{s}(1 - e^{-t_1 s})$$

The response $X_o(s)$ is then obtained as

$$X_o(s) = \frac{k_1(b_2 s + k_2)}{(k_1 + k_2)b_2 s + k_1 k_2} \frac{X_i}{s}(1 - e^{-t_1 s})$$

Since

$$\frac{k_1(b_2 s + k_2)}{(k_1 + k_2)b_2 s + k_1 k_2} \frac{1}{s} = \frac{1}{s} + \frac{-\frac{b_2}{k_1}}{\left(\frac{1}{k_1} + \frac{1}{k_2}\right)b_2 s + 1}$$

we have

$$\mathcal{L}^{-1}\left[\frac{k_1(b_2 s + k_2)}{(k_1 + k_2)b_2 s + k_1 k_2} \frac{1}{s}\right] = 1 - \frac{k_2}{k_1 + k_2} \exp\left\{-t/[(\frac{1}{k_1} + \frac{1}{k_2})b_2]\right\}$$

Hence

$$x_o(t) = X_i\left[1 - \frac{k_2}{k_1 + k_2} \exp\left\{-t/[(\frac{1}{k_1} + \frac{1}{k_2})b_2]\right\}\right]$$
$$- X_i\left[1 - \frac{k_2}{k_1 + k_2} \exp\left\{-(t - t_1)/[(\frac{1}{k_1} + \frac{1}{k_2})b_2]\right\}\right]1(t - t_1)$$

B-8-6. First note that

$$\frac{1}{Z_1} = \frac{1}{R_1} + C_1 s, \qquad Z_2 = R_2 + \frac{1}{C_2 s} = \frac{R_2 C_2 s + 1}{C_2 s}$$

Then

$$\frac{E_o(s)}{E_i(s)} = \frac{Z_2}{Z_1 + Z_2} = \frac{(R_2C_2s + 1)(R_1C_1s + 1)}{R_1C_2s + (R_2C_2s + 1)(R_1C_1s + 1)}$$

or

$$\frac{E_o(s)}{E_i(s)} = \frac{(R_1C_1s + 1)(R_2C_2s + 1)}{R_1C_1R_2C_2s^2 + (R_1C_1 + R_2C_2 + R_1C_2)s + 1}$$

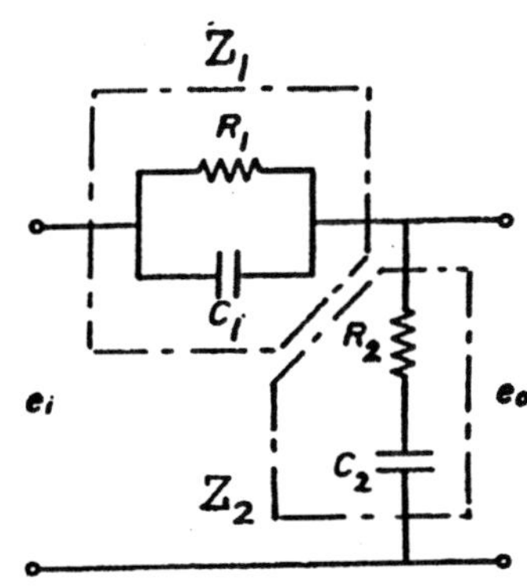

Since $R_2 = 1.5\ R_1$, $C_2 = C_1$, and $R_1C_1 = 1$, we obtain

$$R_1C_2 = R_1C_1 = 1, \qquad R_2C_2 = 1.5\ R_1C_1 = 1.5$$

and the transfer function $E_o(s)/E_i(s)$ becomes

$$\frac{E_o(s)}{E_i(s)} = \frac{(s + 1)(1.5s + 1)}{1.5\ s^2 + 3.5s + 1} = \frac{(s + 1)[s + (2/3)]}{(s + 2)[s + (1/3)]}$$

Since the input $e_i(t)$ is given by

$$e_i(t) = E_i \qquad 0 < t < t_1$$
$$= 0 \qquad \text{elsewhere}$$

we have

$$e_i(t) = E_i[1(t) - 1(t - t_1)]$$

Hence, the response $e_o(t)$ can be obtained as follows:

$$E_o(s) = \frac{(s + 1)[s + (2/3)]}{(s + 2)[s + (1/3)]} E_i \left(\frac{1}{s} - \frac{1}{s} e^{-st_1}\right)$$

Since

$$\mathcal{L}^{-1}\left[\frac{(s + 1)[s + (2/3)]}{(s + 2)[s + (1/3)]s}\right] = \mathcal{L}^{-1}\left[\frac{0.4}{s + 2} - \frac{0.4}{s + (1/3)} + \frac{1}{s}\right]$$

$$= 0.4\ e^{-2t} - 0.4\ e^{-(1/3)t} + 1$$

we obtain the response $e_o(t)$ as follows:

$$e_o(t) = E_i \,(0.4\, e^{-2t} - 0.4\, e^{-(1/3)t} + 1) - E_i\,[0.4\, e^{-2(t - t_1)} - 0.4\, e^{-(t - t_1)/3} + 1] \cdot 1(t - t_1)$$

B-8-7.

$$\text{logarithmic decrement} = \ln \frac{x_0}{x_1} = \frac{1}{n} \ln \frac{x_0}{x_n} = \zeta\omega_n T = \zeta\omega_n \frac{2\pi}{\omega_d} = \frac{2\pi\zeta}{\sqrt{1-\zeta^2}}$$

Thus

$$\frac{1}{n} \ln \frac{x_0}{x_n} = \frac{2\zeta\pi}{\sqrt{1-\zeta^2}}$$

or

$$\left(\frac{1}{n} \ln \frac{x_0}{x_n}\right)^2 (1 - \zeta^2) = 4\zeta^2\pi^2$$

Solving for ζ,

$$\zeta = \frac{\frac{1}{n} \ln \frac{x_0}{x_n}}{\sqrt{4\pi^2 + \left[\frac{1}{n}\left(\ln \frac{x_0}{x_n}\right)\right]^2}}$$

By substituting $n = 4$ and $x_0/x_4 = 4$ into this last equation, we obtain

$$\zeta = \frac{\frac{1}{4} \ln 4}{\sqrt{4\pi^2 + \left[\frac{1}{4}(\ln 4)\right]^2}}$$

$$= \frac{\frac{1}{4} \times 1.386}{\sqrt{39.48 + 0.12}} = \frac{0.3466}{6.293} = 0.055$$

Noting that $\omega_n = \sqrt{k/m}$ and $2\zeta\omega_n = b/m$, we find

$$\omega_n = \sqrt{\frac{k}{m}} = \sqrt{\frac{500}{1}} = 22.36 \text{ rad/s}$$

and

$$b = 2\zeta\omega_n m = 2 \times 0.055 \times 22.36 \times 1 = 2.46 \text{ N-s/m}$$

B-8-8. The system equation is

$$(m + 2)\ddot{x} + b\dot{x} + kx = p$$

Substituting the given numerical values $m = 20$ kg and $p = 2g = 2 \times 9.807$ N into this last equation, we obtain

$$22\,\ddot{x} + b\dot{x} + kx = 2 \times 9.807$$

At steady state

$$kx_{ss} = 2 \times 9.807$$

From Figure 8-28 (b), $x_{ss} = 0.08$ m. Thus

$$x_{ss} = \frac{2 \times 9.807}{k} = 0.08$$

Solving for k, we obtain

$$k = \frac{2 \times 9.807}{0.08} = 245 \text{ N/m}$$

Since

$$\omega_n = \sqrt{\frac{k}{22}} = \sqrt{\frac{245}{22}} = 3.34 \text{ rad/s}, \qquad 2\zeta\omega_n = \frac{b}{22}$$

we obtain

$$b = 2\zeta\omega_n \times 22 = 2 \times 0.4 \times 3.34 \times 22 = 58.8 \text{ N-s/m}$$

B-8-9. For the x direction, the equation of motion is

$$m_1 \frac{d^2}{dt^2} x + m_2 \frac{d^2}{dt^2} (x + \ell \sin \theta) = - 2kx$$

For the rotational motion of the pendulum,

$$m_2 \left[\frac{d^2}{dt^2} (x + \ell \sin \theta) \right] \ell \cos \theta + m_2 \left[\frac{d^2}{dt^2} (-\ell \cos \theta) \right] \ell \sin \theta = - m_2 g \ell \sin \theta$$

Rewriting the preceding two equations,

$$m_1\ddot{x} + m_2(\ddot{x} - \ell \sin\theta\,\dot{\theta}^2 + \ell\cos\theta\,\ddot{\theta}) = -2kx$$

$$m_2(\ddot{x} - \ell\sin\theta\,\dot{\theta}^2 + \ell\cos\theta\,\ddot{\theta})\ell\cos\theta$$
$$+ m_2\ell(\cos\theta\,\dot{\theta}^2 + \sin\theta\,\ddot{\theta})\ell\sin\theta = -m_2 g\ell\sin\theta$$

Simplifying,

$$\ddot{x} + \frac{m_2\ell}{m_1 + m_2}\cos\theta\,\ddot{\theta} - \frac{m_2\ell}{m_1 + m_2}\sin\theta\,\dot{\theta}^2 + \frac{2k}{m_1 + m_2}x = 0$$

$$m_2\ddot{x}\cos\theta + m_2\ell\cos^2\theta\,\ddot{\theta} + m_2\ell\sin^2\theta\,\ddot{\theta} + m_2 g\sin\theta = 0$$

Thus the equations of motion for the system are

$$\ddot{x} + \frac{m_2\ell}{m_1 + m_2}\cos\theta\,\ddot{\theta} - \frac{m_2\ell}{m_1 + m_2}\sin\theta\,\dot{\theta}^2 + \frac{2k}{m_1 + m_2}x = 0$$

$$\ddot{\theta} + \frac{\ddot{x}}{\ell}\cos\theta + \frac{g}{\ell}\sin\theta = 0$$

Next, we shall obtain the response of the system to the given initial condition. We assume that θ is small so that $\cos\theta \doteqdot 1$, $\sin\theta \doteqdot 0$, $\theta^2 \doteqdot 0$ and $\dot{\theta}^2 \doteqdot 0$. By substituting $m_1 = 10$ kg, $m_2 = 1$ kg, $k = 250$ N/m, and $\ell = 1$ m, the equations of the system become as follows:

$$\ddot{x} + \frac{1}{11}\ddot{\theta} + \frac{500}{11}x = 0$$

$$\ddot{\theta} + \ddot{x} + 9.807\,\theta = 0$$

The initial conditions are $x(0) = 0.1$ m, $\dot{x}(0) = 0$ m/s, $\theta(0) = 0$ rad, and $\dot{\theta}(0) = 0$ rad/s.

The Laplace transform of the system equations gives

$$[s^2X(s) - 0.1s] + \frac{1}{11}s^2\Theta(s) + \frac{500}{11}X(s) = 0$$

$$s^2\Theta(s) + [s^2X(s) - 0.1s] + 9.807\Theta(s) = 0$$

Rewriting the last two equations, we get

$$\left(s^2 + \frac{500}{11}\right)X(s) + \frac{1}{11}s^2\Theta(s) = 0.1\,s \tag{1}$$

$$(s^2 + 9.807)\Theta(s) + s^2X(s) = 0.1\,s \tag{2}$$

Subtracting Equation (2) from Equation (1) gives

$$X(s) = 0.02\, s^2\Theta(s) + 0.215754\, \Theta(s) \qquad (3)$$

Substituting Equation (3) into Equation (1) gives

$$(0.02\, s^4 + 1.2158\, s^2 + 9.807)\Theta(s) = 0.1s$$

Hence

$$\Theta(s) = \frac{5s^2}{s^4 + 60.79\, s^2 + 490.35} \frac{1}{s}$$

From Equation (3), we get

$$X(s) = \frac{0.1s^4 + 1.07877\, s^2}{s^4 + 60.79\, s^2 + 490.35} \frac{1}{s}$$

The MATLAB program given below produces the response curves θ rad versus t and x m versus t. The response curves are shown on the next page.

```
>> t = 0:0.01:10;
>> num1 = [5    0    0];
>> num2 = [0.1    0    1.07877    0    0];
>> den = [1    0    60.79    0    490.35];
>> z = step(num1,den,t);
>> x = step(num2,den,t);
>> subplot(221); plot(t,z)
>> grid
>> title('Response to Initial Condition')
>> xlabel('t (sec)'); ylabel('\theta')
>> subplot(222); plot(t,x)
>> grid
>> title('Response to Initial Condition')
>> xlabel('t (sec)'); ylabel('x')
>>
>> roots(den)

ans =

        0 + 7.1565i
        0 - 7.1565i
        0 + 3.0942i
        0 - 3.0942i
```

Notice that the roots of the denominator polynomial gives the natural frequencies of the system. They are $_1$ = 7.1565 rad/s and $_2$ = 3.0942 rad/s.

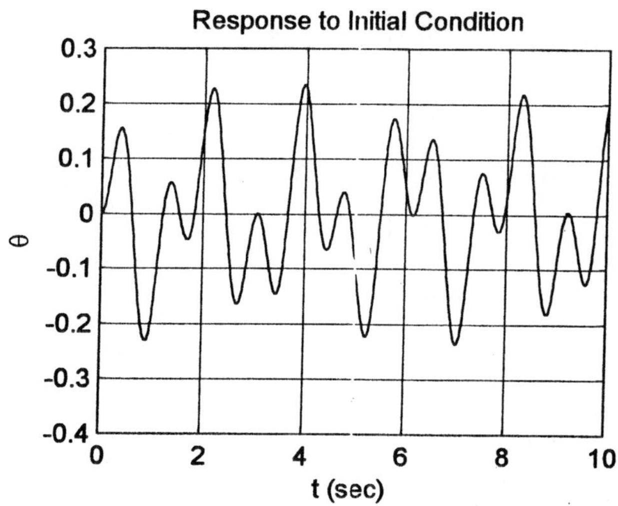

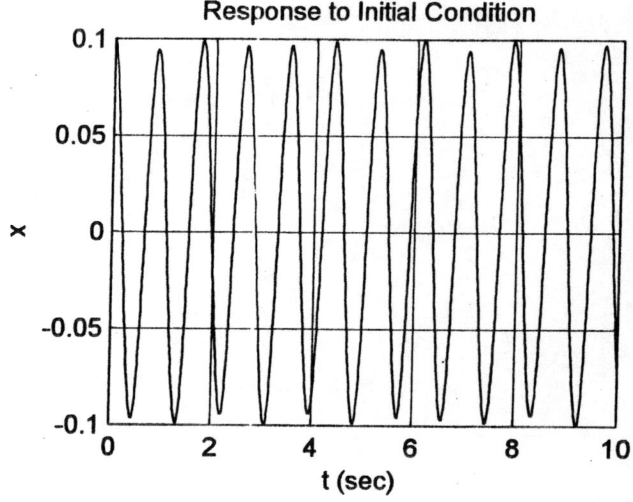

B-8-10. The equation of motion for the system is

$$m\ddot{x} + kx = p(t)$$

where m = 1 kg, k = 100 N/m, $p(t) = 10\ \delta(t)$ N, $x(0-) = 0.1$ m, and $\dot{x}(0-) = 1$ m/s. By substituting the given numerical values into the system equation, we obtain

$$\ddot{x} + 100\ x = 10\ \delta(t)$$

Taking the $\mathcal{L}_-$ transform of this last equation gives

$$[s^2X(s) - sx(0-) - \dot{x}(0-)] + 100\ X(s) = 10$$

or

$$(s^2 + 100)X(s) = 10 + 0.1s + 1 = 11 + 0.1s$$

Solving for X(s) gives

$$X(s) = \frac{11 + 0.1s}{s^2 + 10^2}$$

The inverse Laplace transform of X(s) gives

$$x(t) = \frac{11}{10} \sin 10t + 0.1 \cos 10t$$

B-8-11. The system equation is

$$m\ddot{x} + b\dot{x} = \delta(t), \qquad x(0-) = 0, \qquad \dot{x}(0-) = 0$$

The $\mathcal{L}_-$ transform of this equation is

$$(ms^2 + bs)X(s) = 1$$

Solving for X(s), we get

$$X(s) = \frac{1}{ms^2 + bs} = \frac{1}{m} \frac{1}{s(s + \frac{b}{m})} = \frac{1}{b}\left(\frac{1}{s} - \frac{1}{s + \frac{b}{m}}\right)$$

The response x(t) of the system is

$$x(t) = \frac{1}{b} [1 - e^{-(b/m)t}]$$

The velocity $\dot{x}(t)$ is

$$\dot{x}(t) = \frac{1}{m} e^{-(b/m)t}$$

Thus

$$\dot{x}(0) = \frac{1}{m}$$

The initial velocity can also be obtained by use of the initial value theorem.

$$\dot{x}(0+) = \lim_{s\to\infty} s^2X(s) = \lim_{s\to\infty} \frac{s^2}{ms^2 + bs} = \frac{1}{m}$$

B-8-12. In this system u is the input and x_2 is the output. From Figure 8-32, we obtain the following equations:

$$b_1\dot{x}_1 + k_1(x_1 - x_2) = u$$

$$k_1(x_1 - x_2) = b_2\dot{x}_2 + k_2x_2$$

Laplace transforming these two equations, assuming zero initial conditions, we obtain

$$(b_1s + k_1)X_1(s) - k_1X_2(s) = U(s)$$

$$k_1X_1(s) = (b_2s + k_1 + k_2)X_2(s)$$

By eliminating $X_1(s)$ from these two equations, we get

$$(b_1s + k_1) \frac{b_2s + k_1 + k_2}{k_1} X_2(s) - k_1X_2(s) = U(s)$$

Simplifying this last equation, we get

$$[b_1b_2s^2 + (b_1k_1 + b_1k_2 + b_2k_1)s + k_1k_2]X_2(s) = k_1U(s)$$

from which we obtain

$$\frac{X_2(s)}{U(s)} = \frac{k_1}{b_1 b_2 s^2 + (b_1 k_1 + b_1 k_2 + b_2 k_1)s + k_1 k_2}$$

By substituting numerical values for k_1, k_2, b_1, and b_2 into this last equation, we obtain

$$\frac{X_2(s)}{U(s)} = \frac{4}{10s^2 + (4 + 20 + 40)s + 4 \times 20}$$

$$= \frac{0.4}{s^2 + 6.4s + 8}$$

Since the input U is a step force of 2N, we have U(s) = 2/s. $X_2(s)$ can be obtained from

$$X_2(s) = \frac{0.4}{s^2 + 6.4s + 8} \frac{2}{s}$$

$$= \frac{0.8}{(s + 4.6967)(s + 1.7033)s}$$

$$= \frac{0.1}{s} + \frac{0.0569}{s + 4.6967} + \frac{-0.1569}{s + 1.7033}$$

The inverse Laplace transform of $X_2(s)$ gives

$$x_2(t) = 0.1 + 0.0569\ e^{-4.6967t} - 0.1569\ e^{-1.7033t}$$

The following MATLAB program produces the step-response curve. The resulting step-response curve $x_2(t)$ versus t is shown below.

```
>> num = [0    0    0.8];
>> den = [1    6.4    8];
>> t = 0:0.01:5;
>> y = step(num,den,t);
>> plot(t,y)
>> grid
>> title('Step Response')
>> xlabel('t (sec)')
>> ylabel('x_2(t)')
```

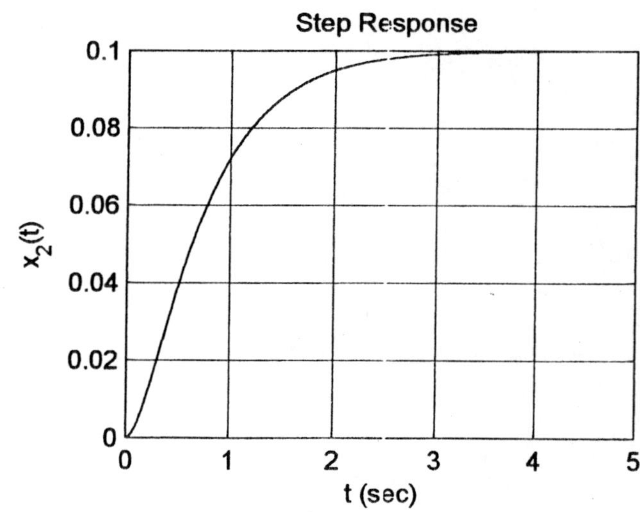

B-8-13. Referring to the figure shown to the right, we have

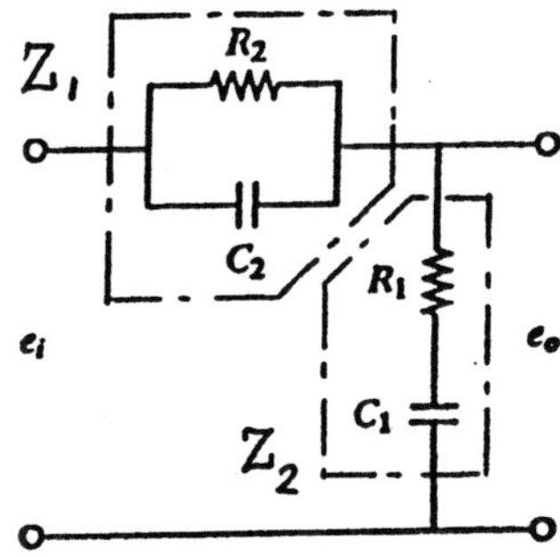

$$\frac{1}{Z_1(s)} = \frac{1}{R_2} + C_2 s, \qquad Z_2(s) = R_1 + \frac{1}{C_1 s}$$

Hence

$$Z_1(s) = \frac{R_2}{R_2 C_2 s + 1}, \qquad Z_2(s) = \frac{R_1 C_1 s + 1}{C_1 s}$$

Thus

$$\frac{E_o(s)}{E_i(s)} = \frac{Z_2(s)}{Z_1(s) + Z_2(s)}$$

$$= \frac{(R_1 C_1 s + 1)(R_2 C_2 s + 1)}{R_2 C_1 s + (R_1 C_1 s + 1)(R_2 C_2 s + 1)}$$

$$= \frac{(R_1 C_1 s + 1)(R_2 C_2 s + 1)}{R_1 C_1 R_2 C_2 s^2 + (R_1 C_1 + R_2 C_2 + R_2 C_1)s + 1}$$

By substituting the given numerical values for R_1, R_2, C_1, and C_2, we obtain

$$\frac{E_o(s)}{E_i(s)} = \frac{(0.5s + 1)(0.05s + 1)}{0.025s^2 + 0.8s + 1}$$

When $e_i(t) = 5$ V (step input) is applied to the system, we have

$$E_o(s) = \frac{0.025s^2 + 0.55s + 1}{0.025s^2 + 0.8s + 1} \frac{5}{s}$$

$$= \frac{5s^2 + 110s + 200}{s^2 + 32s + 40} \frac{1}{s}$$

The response curve $e_o(t)$ versus t can be obtained by entering the following MATLAB program into the computer.

```
>> num = [5   110   200];
>> den = [1   32   40];
>> t = 0:0.01:5;
>> y = step(num,den,t);
>> subplot(221); plot(t,y)
>> v = [-1   5   0   6]; axis(v);
>> grid
>> title('Step Response')
>> xlabel('t (sec)')
>> ylabel('Output e_o(t)')
```

The resulting response curve $e_o(t)$ versus t is shown below.

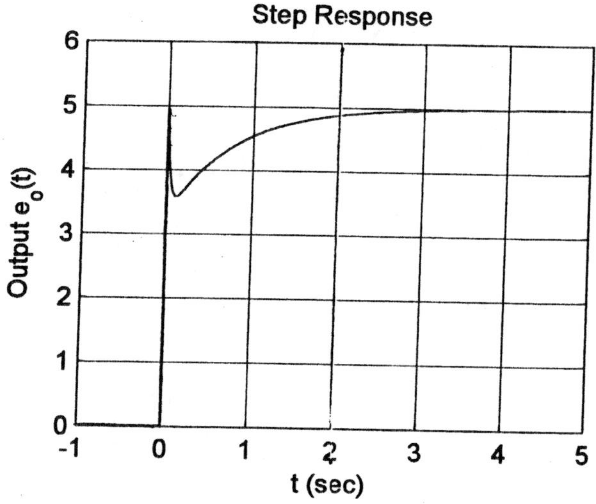

Note that

$$E_o(s) = \frac{5s^2 + 110s + 200}{s^2 + 32s + 40} \frac{1}{s}$$

$$= \frac{1.7010}{s + 30.6969} - \frac{1.7010}{s + 1.3031} + \frac{5}{s}$$

Hence

$$e_o(t) = 1.7010(e^{-30.6969t} - e^{-1.3031t}) + 5$$

Notice that the response curve is a sum of two exponential curves and a step function of magnitude 5.

B-8-14. Note that the Laplace transform of $u(t) = \frac{1}{2}t^2$ is

$$U(s) = \frac{1}{s^3}$$

Hence

$$Y(s) = \frac{1}{s^2 + s + 1} \frac{1}{s^3} = \frac{1}{s^5 + s^4 + s^3} \qquad (1)$$

A partial-fraction expansion of Y(s) can be obtained by use of MATLAB. See the MATLAB program shown on the next page.

```
>> num = [1]; den = [1   1   1   0   0   0];
>> [r,p,K] = residue(num,den)

r =

  -0.0000 - 0.5774i
  -0.0000 + 0.5774i
   0.0000
  -1.0000
   1.0000


p =

  -0.5000 + 0.8660i
  -0.5000 - 0.8660i
        0
        0
        0


K =

     []
```

From the MATLAB output we obtain

$$Y(s) = \frac{-0.5774j}{s + 0.5 + 0.866j} + \frac{0.5774j}{s + 0.5 - 0.866j} + \frac{0}{s} - \frac{1}{s^2} + \frac{1}{s^3}$$

$$= \frac{-1}{s^2 + s + 1} - \frac{1}{s^2} + \frac{1}{s^3}$$

The inverse Laplace transform of Y(s) can be obtained as follows:

$$y(t) = \frac{1}{0.866} e^{-0.5t} \sin 0.866t - t + \tfrac{1}{2} t^2$$

A MATLAB program for obtaining the response y(t) [the inverse Laplace transform of Y(s) given by Equation (1)] by use of a step command is shown on the next page. The resulting response curve y(t) versus t is also shown on the next page.

```
>> num = [1   0];
>> den = [1   1   1   0   0   0];
>> t = 0:0.1:20;
>> y = step(num,den,t);
>> plot(t,y)
>> grid
>> title ('Response to 0.5*t^2')
>> xlabel('t (sec)')
>> ylabel('Output y(t)')
```

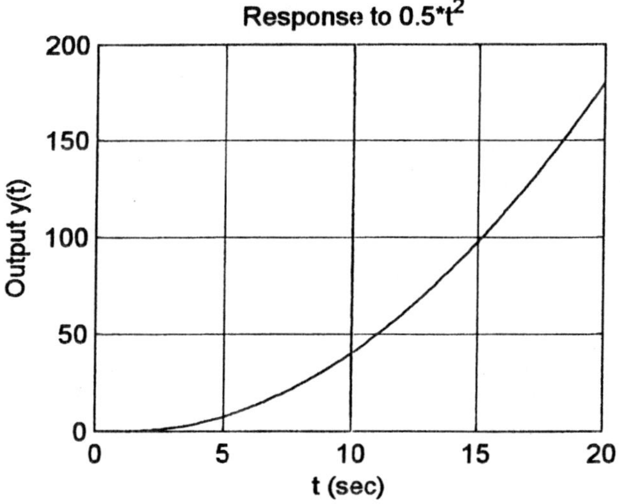

B-8-15. There are several different ways to plot unit-impulse response curves of

$$\frac{Y(s)}{U(s)} = \frac{1}{s^2 + 2\zeta s + 1}$$

for different values of ζ. One way to plot these response curves is to use a "for" loop as shown in the MATLAB program given in the next page. The program produces both two-dimensional and three-dimensional impulse response curves.

```
>> % --- Two-dimensional plot ---
>>
>> t = 0:0.2:10;
>> zeta = [0   0.2   0.4   0.6   0.8   1];
>>       for n = 1:6;
             num = [0   0   1];
             den = [1   2*zeta(n)    1];
             [y(1:51,n),x,t] = impulse(num,den,t);
       end
>> subplot(221); plot(t,y)
>> grid
>> title('Two-Dim Plot of Unit-Impulse Response Curves')
>> xlabel('t (sec)')
>> ylabel('Response')
>> text(3.3,0.90,'\zeta = 0')
>> text(3.3,0.75,'0.2')
>> text(3.3,0.65,'0.4')
>> text(3.3,0.55,'0.6')
>> text(3.3,0.45,'0.8')
>> text(3.3,0.35,'1.0')
>>
>> % --- Three-dimensional plot ---
>>
>> subplot(222); mesh(t,zeta,y')
>> title('Three-Dim Plot of Unit-Impulse Response Curves')
>> xlabel('t (sec)')
>> ylabel('\zeta')
>> zlabel('Response')
```

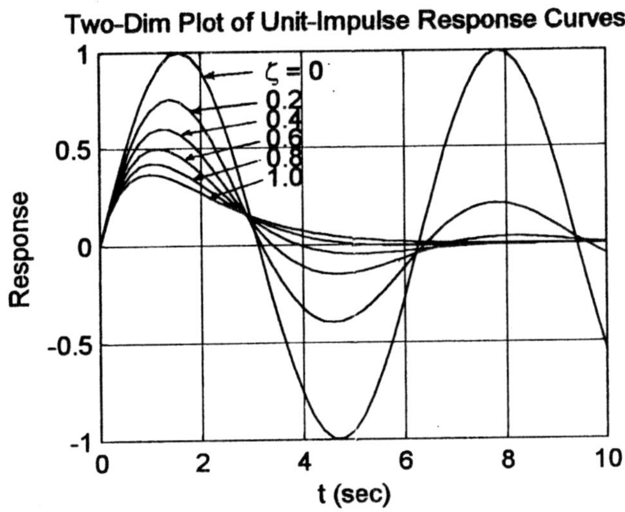

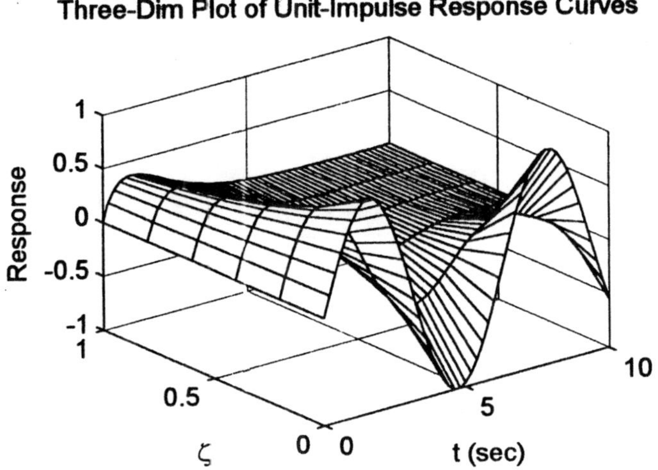

B-8-16.

$$e^{\mathbf{A}t} = \mathcal{L}^{-1}[(s\mathbf{I} - \mathbf{A})^{-1}] = \mathcal{L}^{-1}\left\{\begin{bmatrix} s+2 & 1 \\ -2 & s+5 \end{bmatrix}^{-1}\right\}$$

$$= \mathcal{L}^{-1}\left\{\frac{1}{s^2 + 7s + 12}\begin{bmatrix} s+5 & -1 \\ 2 & s+2 \end{bmatrix}\right\}$$

$$= \mathcal{L}^{-1}\begin{bmatrix} \dfrac{s+5}{(s+3)(s+4)} & \dfrac{-1}{(s+3)(s+4)} \\ \dfrac{2}{(s+3)(s+4)} & \dfrac{s+2}{(s+3)(s+4)} \end{bmatrix}$$

$$= \mathcal{L}^{-1}\begin{bmatrix} \dfrac{2}{s+3} - \dfrac{1}{s+4} & -\dfrac{1}{s+3} + \dfrac{1}{s+4} \\ \dfrac{2}{s+3} - \dfrac{2}{s+4} & -\dfrac{1}{s+3} + \dfrac{2}{s+4} \end{bmatrix}$$

$$= \begin{bmatrix} 2e^{-3t} - e^{-4t} & -e^{-3t} + e^{-4t} \\ 2e^{-3t} - 2e^{-4t} & -e^{-3t} + 2e^{-4t} \end{bmatrix}$$

B-8-17.

$$e^{\mathbf{A}t} = \mathcal{L}^{-1}[(s\mathbf{I} - \mathbf{A})^{-1}]$$

$$= \mathcal{L}^{-1}\left\{\begin{bmatrix} s & -1 & 0 \\ 0 & s & -1 \\ -1 & 3 & s-3 \end{bmatrix}^{-1}\right\}$$

where

$$\begin{bmatrix} s & -1 & 0 \\ 0 & s & -1 \\ -1 & 3 & s-3 \end{bmatrix}^{-1}$$

$$= \frac{1}{s^2(s-3) - 1 + 3s} \begin{bmatrix} \begin{vmatrix} s & -1 \\ 3 & s-3 \end{vmatrix} & -\begin{vmatrix} -1 & 0 \\ 3 & s-3 \end{vmatrix} & \begin{vmatrix} -1 & 0 \\ s & -1 \end{vmatrix} \\ -\begin{vmatrix} 0 & -1 \\ -1 & s-3 \end{vmatrix} & \begin{vmatrix} s & 0 \\ -1 & s-3 \end{vmatrix} & -\begin{vmatrix} s & 0 \\ 0 & -1 \end{vmatrix} \\ \begin{vmatrix} 0 & s \\ -1 & 3 \end{vmatrix} & -\begin{vmatrix} s & -1 \\ -1 & 3 \end{vmatrix} & \begin{vmatrix} s & -1 \\ 0 & s \end{vmatrix} \end{bmatrix}$$

$$= \frac{1}{(s-1)^3} \begin{bmatrix} s^2 - 3s + 3 & s - 3 & 1 \\ 1 & s(s-3) & s \\ s & -(3s-1) & s^2 \end{bmatrix}$$

$$= \begin{bmatrix} \dfrac{s^2 - 3s + 3}{(s-1)^3} & \dfrac{s-3}{(s-1)^3} & \dfrac{1}{(s-1)^3} \\ \dfrac{1}{(s-1)^3} & \dfrac{s(s-3)}{(s-1)^3} & \dfrac{s}{(s-1)^3} \\ \dfrac{s}{(s-1)^3} & \dfrac{-3s+1}{(s-1)^3} & \dfrac{s^2}{(s-1)^3} \end{bmatrix}$$

$$= \begin{bmatrix} \dfrac{1}{s-1} - \dfrac{1}{(s-1)^2} + \dfrac{1}{(s-1)^3} & \dfrac{1}{(s-1)^2} - \dfrac{2}{(s-1)^3} & \dfrac{1}{(s-1)^3} \\ \dfrac{1}{(s-1)^3} & \dfrac{1}{s-1} - \dfrac{1}{(s-1)^2} - \dfrac{2}{(s-1)^3} & \dfrac{1}{(s-1)^2} + \dfrac{1}{(s-1)^3} \\ \dfrac{1}{(s-1)^2} + \dfrac{1}{(s-1)^3} & -\dfrac{3}{(s-1)^2} - \dfrac{2}{(s-1)^3} & \dfrac{1}{s-1} + \dfrac{2}{(s-1)^2} + \dfrac{1}{(s-1)^3} \end{bmatrix}$$

Noting that

$$\mathcal{L}^{-1}\left[\frac{1}{s-1}\right] = e^t$$

$$\mathcal{L}^{-1}\left[\frac{1}{(s-1)^2}\right] = te^t$$

$$\mathcal{L}^{-1}\left[\frac{1}{(s-1)^3}\right] = \frac{1}{2} t^2 e^t$$

we obtain

$$e^{\mathbf{A}t} = \mathcal{L}^{-1}\left\{\begin{bmatrix} s & -1 & 0 \\ 0 & s & -1 \\ -1 & 3 & s-3 \end{bmatrix}^{-1}\right\}$$

$$= \begin{bmatrix} e^t - te^t + \frac{1}{2}t^2e^t & te^t - t^2e^t & \frac{1}{2}t^2e^t \\ \frac{1}{2}t^2e^t & e^t - te^t - t^2e^t & te^t + \frac{1}{2}t^2e^t \\ te^t + \frac{1}{2}t^2e^t & -3te^t - t^2e^t & e^t + 2te^t + \frac{1}{2}t^2e^t \end{bmatrix}$$

B-8-18. The solution of the state equation is

$$\mathbf{x}(t) = e^{\mathbf{A}t}\mathbf{x}(0) + \int_0^t e^{\mathbf{A}(t-\tau)}\,\mathbf{B}u(\tau)\,d\tau$$

Since $\mathbf{x}(0) = \mathbf{0}$ and $u(t) = 1(t)$, we obtain

$$\mathbf{x}(t) = \int_0^t e^{\mathbf{A}(t-\tau)}\,\mathbf{B}\,d\tau$$

Note that

$$e^{\mathbf{A}t} = \mathcal{L}^{-1}[(s\mathbf{I} - \mathbf{A})^{-1}]$$

where

$$(s\mathbf{I} - \mathbf{A})^{-1} = \begin{bmatrix} s & -1 \\ 6 & s+5 \end{bmatrix}^{-1} = \frac{1}{s^2 + 5s + 6}\begin{bmatrix} s+5 & 1 \\ -6 & s \end{bmatrix}$$

$$= \begin{bmatrix} \dfrac{s+5}{(s+2)(s+3)} & \dfrac{1}{(s+2)(s+3)} \\ \dfrac{-6}{(s+2)(s+3)} & \dfrac{s}{(s+2)(s+3)} \end{bmatrix}$$

$$= \begin{bmatrix} \dfrac{3}{s+2} - \dfrac{2}{s+3} & \dfrac{1}{s+2} - \dfrac{1}{s+3} \\ -\dfrac{6}{s+2} + \dfrac{6}{s+3} & -\dfrac{2}{s+2} + \dfrac{3}{s+3} \end{bmatrix}$$

Thus

$$e^{\underset{\sim}{A}t} = \mathcal{L}^{-1}[(s\underset{\sim}{I} - \underset{\sim}{A})^{-1}]$$

$$= \begin{bmatrix} 3\,e^{-2t} - 2\,e^{-3t} & e^{-2t} - e^{-3t} \\ -6\,e^{-2t} + 6\,e^{-3t} & -2\,e^{-2t} + 3\,e^{-3t} \end{bmatrix}$$

Hence

$$x(t) = \int_0^t e^{\underset{\sim}{A}(t-\tau)}\,\underset{\sim}{B}\,d\tau = e^{\underset{\sim}{A}t}\int_0^t e^{-\underset{\sim}{A}\tau}\begin{bmatrix} 0 \\ 1 \end{bmatrix} d\tau$$

$$= \begin{bmatrix} 3\,e^{-2t} - 2\,e^{-3t} & e^{-2t} - e^{-3t} \\ -6\,e^{-2t} + 6\,e^{-3t} & -2\,e^{-2t} + 3\,e^{-3t} \end{bmatrix} \int_0^t \begin{bmatrix} e^{2\tau} - e^{3\tau} \\ -2\,e^{2\tau} + 3\,e^{3\tau} \end{bmatrix} d\tau$$

$$= \begin{bmatrix} 3\,e^{-2t} - 2\,e^{-3t} & e^{-2t} - e^{-3t} \\ -6\,e^{-2t} + 6\,e^{-3t} & -2\,e^{-2t} + 3\,e^{-3t} \end{bmatrix} \begin{bmatrix} \frac{1}{2}\,e^{2t} - \frac{1}{3}\,e^{3t} - \frac{1}{6} \\ -\,e^{2t} + e^{3t} \end{bmatrix}$$

$$= \begin{bmatrix} \frac{1}{6} - \frac{1}{2}\,e^{-2t} + \frac{1}{3}\,e^{-3t} \\ e^{-2t} - e^{-3t} \end{bmatrix}$$

<u>Alternate solution</u> Referring to Equation (8-39), we have for the unit-step input the following solution:

$$\underset{\sim}{x}(t) = e^{\underset{\sim}{A}t}\,\underset{\sim}{x}(0) + \underset{\sim}{A}^{-1}(e^{\underset{\sim}{A}t} - \underset{\sim}{I})\underset{\sim}{B}$$

Since $\underset{\sim}{x}(0) = \underset{\sim}{0}$ in the present case, we have

$$\underset{\sim}{x}(t) = \underset{\sim}{A}^{-1}(e^{\underset{\sim}{A}t} - \underset{\sim}{I})\underset{\sim}{B}$$

where

$$\underset{\sim}{A}^{-1} = \begin{bmatrix} 0 & 1 \\ -6 & -5 \end{bmatrix}^{-1} = \begin{bmatrix} -\frac{5}{6} & -\frac{1}{6} \\ 1 & 0 \end{bmatrix}$$

and

$$e^{\underset{\sim}{A}t} = \begin{bmatrix} 3\,e^{-2t} - 2\,e^{-3t} & e^{-2t} - e^{-3t} \\ -6\,e^{-2t} + 6\,e^{-3t} & -2\,e^{-2t} + 3\,e^{-3t} \end{bmatrix}$$

Thus,

$$(e^{\underset{\sim}{A}t} - \underset{\sim}{I})\underset{\sim}{B} = \begin{bmatrix} 3\,e^{-2t} - 2\,e^{-3t} - 1 & e^{-2t} - e^{-3t} \\ -6\,e^{-2t} + 6\,e^{-3t} & -2\,e^{-2t} + 3\,e^{-3t} - 1 \end{bmatrix} \begin{bmatrix} 0 \\ 1 \end{bmatrix}$$

$$= \begin{bmatrix} e^{-2t} - e^{-3t} \\ -2\,e^{-2t} + 3\,e^{-3t} - 1 \end{bmatrix}$$

and $\underset{\sim}{x}(t)$ is given by

$$\underset{\sim}{x}(t) = \begin{bmatrix} -\frac{5}{6} & -\frac{1}{6} \\ 1 & 0 \end{bmatrix} \begin{bmatrix} e^{-2t} - e^{-3t} \\ -2\,e^{-2t} + 3\,e^{-3t} - 1 \end{bmatrix}$$

$$= \begin{bmatrix} \frac{1}{6} - \frac{1}{2}\,e^{-2t} + \frac{1}{3}\,e^{-3t} \\ e^{-2t} - e^{-3t} \end{bmatrix}$$

B-8-19. For the given system

$$\begin{bmatrix} \dot{x}_1 \\ \dot{x}_2 \end{bmatrix} = \begin{bmatrix} 0 & 1 \\ -6 & -5 \end{bmatrix} \begin{bmatrix} x_1 \\ x_2 \end{bmatrix} + \begin{bmatrix} 0 & 1 \\ 1 & 1 \end{bmatrix} \begin{bmatrix} u_1 \\ u_2 \end{bmatrix}$$

where

$$\begin{bmatrix} u_1 \\ u_2 \end{bmatrix} = \begin{bmatrix} 2 \cdot 1(t) \\ 5 \cdot 1(t) \end{bmatrix} = \begin{bmatrix} 2 \\ 5 \end{bmatrix} 1(t) = \mathbf{k} \cdot 1(t), \qquad \begin{bmatrix} x_1(0) \\ x_2(0) \end{bmatrix} = \begin{bmatrix} 0 \\ 0 \end{bmatrix}$$

Referring to Problem A-8-13, for the system defined by

$$\dot{\mathbf{x}} = \mathbf{A}\mathbf{x} + \mathbf{B}\mathbf{u}$$

we have the following solution to the step input $\mathbf{u} = \mathbf{k} \cdot 1(t)$:

$$\mathbf{x}(t) = e^{\mathbf{A}t}\mathbf{x}(0) + \mathbf{A}^{-1}(e^{\mathbf{A}t} - \mathbf{I})\mathbf{B}\mathbf{k} \qquad (1)$$

For this problem $\mathbf{x}(0) = \mathbf{0}$. Hence, Equation (1) becomes

$$\mathbf{x}(t) = \mathbf{A}^{-1}(e^{\mathbf{A}t} - \mathbf{I})\mathbf{B}\mathbf{k} \qquad (2)$$

Substituting

$$\mathbf{A}^{-1} = \begin{bmatrix} 0 & 1 \\ -6 & -5 \end{bmatrix}^{-1} = \begin{bmatrix} -\frac{5}{6} & -\frac{1}{6} \\ 1 & 0 \end{bmatrix}$$

and

$$\mathbf{B}\mathbf{k} = \begin{bmatrix} 0 & 1 \\ 1 & 1 \end{bmatrix} \begin{bmatrix} 2 \\ 5 \end{bmatrix} = \begin{bmatrix} 5 \\ 7 \end{bmatrix}$$

into Equation (2), we obtain

$$\mathbf{x}(t) = \begin{bmatrix} -\frac{5}{6} & -\frac{1}{6} \\ 1 & 0 \end{bmatrix} (e^{\mathbf{A}t} - \mathbf{I}) \begin{bmatrix} 5 \\ 7 \end{bmatrix} \qquad (3)$$

Referring to the solution to Problem B-8-18, we have

$$e^{\mathbf{A}t} - \mathbf{I} = \begin{bmatrix} 3e^{-2t} - 2e^{-3t} - 1 & e^{-2t} - e^{-3t} \\ -6e^{-2t} + 6e^{-3t} & -2e^{-2t} + 3e^{-3t} - 1 \end{bmatrix} \qquad (4)$$

By substituting Equation (4) into Equation (3) and simplifying, we get

$$\mathbf{x}(t) = \begin{bmatrix} -\frac{5}{6} & -\frac{1}{6} \\ 1 & 0 \end{bmatrix} \begin{bmatrix} 15e^{-2t} - 10e^{-3t} - 5 + 7e^{-2t} - 7e^{-3t} \\ -30e^{-2t} + 30e^{-3t} - 14e^{-2t} + 21e^{-3t} - 7 \end{bmatrix}$$

$$= \begin{bmatrix} -\frac{5}{6} & -\frac{1}{6} \\ 1 & 0 \end{bmatrix} \begin{bmatrix} 22e^{-2t} - 17e^{-3t} - 5 \\ -44e^{-2t} + 51e^{-3t} - 7 \end{bmatrix}$$

$$= \begin{bmatrix} -11e^{-2t} + \frac{17}{3} e^{-3t} + \frac{16}{3} \\ 22e^{-2t} - 17e^{-3t} - 5 \end{bmatrix}$$

This is the solution x(t) obtained analytically.

Notice that

$$\mathbf{x}(0) = \begin{bmatrix} -11 + \frac{17}{3} + \frac{16}{3} \\ 22 - 17 - 5 \end{bmatrix} = \begin{bmatrix} 0 \\ 0 \end{bmatrix} \qquad \mathbf{x}(\infty) = \begin{bmatrix} 5.3333 \\ -5 \end{bmatrix}$$

B-8-20. Notice that for the given system

$$\mathbf{Bu} = \begin{bmatrix} 0 & 1 \\ 1 & 1 \end{bmatrix} \begin{bmatrix} u_1 \\ u_2 \end{bmatrix} = \begin{bmatrix} 0 & 1 \\ 1 & 1 \end{bmatrix} \begin{bmatrix} 2 \\ 5 \end{bmatrix} 1(t) = \begin{bmatrix} 5 \\ 7 \end{bmatrix} 1(t)$$

Hence the state-space representation of the system is given by

$$\begin{bmatrix} \dot{x}_1 \\ \dot{x}_2 \end{bmatrix} = \begin{bmatrix} 0 & 1 \\ -6 & -5 \end{bmatrix} \begin{bmatrix} x_1 \\ x_2 \end{bmatrix} + \begin{bmatrix} 5 \\ 7 \end{bmatrix} 1(t)$$

$$\begin{bmatrix} y_1 \\ y_2 \end{bmatrix} = \begin{bmatrix} 1 & 0 \\ 0 & 1 \end{bmatrix} \begin{bmatrix} x_1 \\ x_2 \end{bmatrix} + \begin{bmatrix} 0 \\ 0 \end{bmatrix} 1(t)$$

The MATLAB program shown below produces the outputs $x_1(t)$ and $x_2(t)$ when the input is the step input. The resulting curves $x_1(t)$ versus t and $x_2(t)$ versus t are shown below the MATLAB program.

```
>> A = [0   1;-6   -5];
>> B = [5;7];
>> C = [1   0;0   1];
>> D = [0;0];
>> t = 0:0.01:8;
>> y = step(A,B,C,D,1,t);
>> x1 = [1   0]*y';
>> x2 = [0   1]*y';
>> subplot(221); plot(t,x1)
>> grid
>> title('Step Response')
>> xlabel('t (sec)')
>> ylabel('x_1')
>> subplot(222); plot(t,x2)
>> grid
>> title('Step Response')
>> xlabel('t (sec)')
>> ylabel('x_2')
```

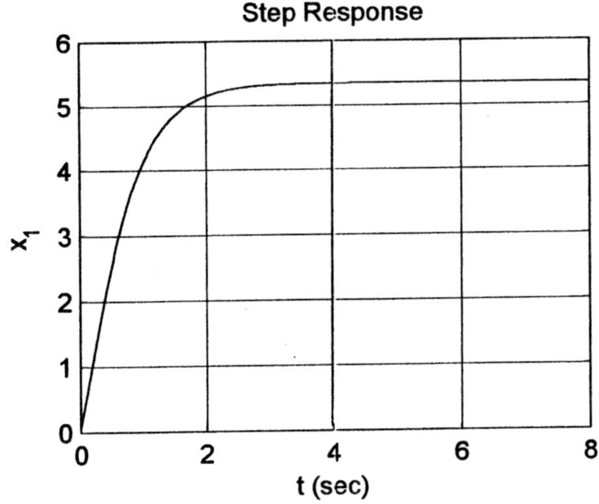

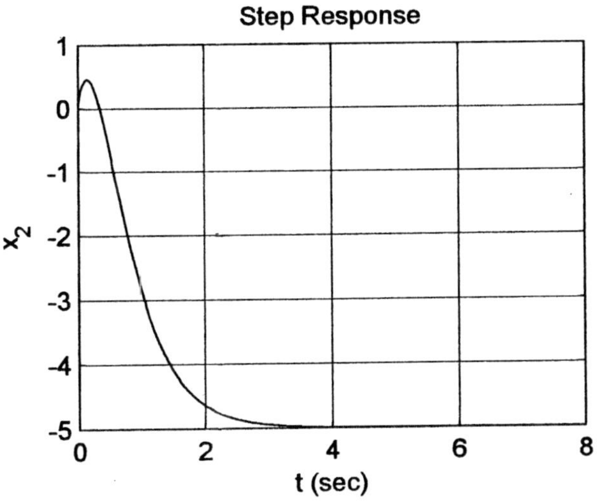

CHAPTER 9

B-9-1. The equation of motion for the system is

$$m\ddot{x} + kx = p(t) = P \sin \omega t$$

where $x(0) = 0$ and $\dot{x}(0) = 0$. The Laplace transform of this equation is

$$(ms^2 + k)X(s) = P \frac{\omega}{s^2 + \omega^2}$$

By substituting the given numerical values of m, k, P, and ω into this last equation, we obtain

$$(s^2 + 100)X(s) = \frac{5 \times 2}{s^2 + 2^2}$$

Solving for X(s),

$$X(s) = \frac{10}{(s^2 + 100)(s^2 + 4)} = \frac{5}{96} \frac{2}{s^2 + 4} - \frac{1}{96} \frac{10}{s^2 + 100}$$

The inverse Laplace transform of X(s) gives the response x(t).

$$x(t) = \frac{5}{96} (\sin 2t - \frac{1}{5} \sin 10t)$$

B-9-2. The equation of motion for the system is

$$m\ddot{x} + b\dot{x} + kx = p(t) = P \sin \omega t, \qquad x(0) = 0, \qquad \dot{x}(0) = 0$$

By substituting the given numerical values into this equation, we get

$$2\ddot{x} + 24\dot{x} + 200x = 5 \sin 6t$$

or

$$\ddot{x} + 12\dot{x} + 100x = 2.5 \sin 6t$$

Taking the Laplace transform of this last equation, we obtain

$$(s^2 + 12s + 100)X(s) = 2.5 \frac{6}{s^2 + 6^2}$$

Solving for X(s),

$$X(s) = \frac{15}{(s^2 + 12s + 100)(s^2 + 36)}$$

$$= \frac{\frac{9}{464}s + \frac{15}{116}}{s^2 + 12s + 100} + \frac{-\frac{9}{464}s + \frac{3}{29}}{s^2 + 36}$$

$$= \frac{\frac{3}{232 \times 8} \times 8 + \frac{9}{464}(s + 6)}{(s + 6)^2 + 8^2} + \frac{3}{29 \times 6}\frac{6}{s^2 + 36}$$

$$- \frac{9}{464}\frac{s}{s^2 + 36}$$

The inverse Laplace transform of X(s) gives

$$x(t) = \frac{3}{1856}e^{-6t}\sin 8t + \frac{9}{464}e^{-6t}\cos 8t + \frac{1}{58}\sin 6t - \frac{9}{464}\cos 6t$$

B-9-3.

$$\frac{E_o(s)}{E_i(s)} = \frac{\frac{1}{Cs}}{R + \frac{1}{Cs}} = \frac{1}{RCs + 1}$$

Hence

$$G(j\omega) = \frac{E_o(j\omega)}{E_i(j\omega)} = \frac{1}{RCj\omega + 1}$$

Thus

$$\left|G(j\omega)\right| = \frac{1}{\sqrt{R^2C^2\omega^2 + 1}}, \quad \angle G(j\omega) = -\tan^{-1} RC\omega$$

For the input $e_i(t) = E_i \sin \omega t$, the steady-state output $e_o(t)$ is given by

$$e_o(t) = \frac{E_i}{\sqrt{R^2C^2\omega^2 + 1}} \sin(\omega t - \tan^{-1} RC\omega)$$

B-9-4. The equations of motion for the system are

$$m_1\ddot{x}_1 + b(\dot{x}_1 - \dot{x}_2) + kx_1 + k(x_1 - x_2) = p(t)$$

$$m_2\ddot{x}_2 + b(\dot{x}_2 - \dot{x}_1) + k(x_2 - x_1) = 0$$

which can be rewritten as

$$m_1\ddot{x}_1 + b\dot{x}_1 + 2kx_1 = b\dot{x}_2 + kx_2 + p(t)$$

$$m_2\ddot{x}_2 + b\dot{x}_2 + kx_2 = b\dot{x}_1 + kx_1$$

Since we are interested in the steady state behavior of the system, we can assume that all initial conditions are zero. Thus, by assuming zero initial conditions and taking the Laplace transforms of the last two equations, we obtain

$$(m_1 s^2 + bs + 2k)X_1(s) = (bs + k)X_2(s) + P(s) \tag{1}$$

$$(m_2 s^2 + bs + k)X_2(s) = (bs + k)X_1(s) \tag{2}$$

From Equation (2) we have

$$X_2(s) = \frac{bs + k}{m_2 s^2 + bs + k} X_1(s) \tag{3}$$

Substituting Equation (3) into Equation (1) and simplifying, we obtain

$$[(m_1 s^2 + k)(m_2 s^2 + bs + k) + (bs + k)m_2 s^2]X_1(s)$$
$$= (m_2 s^2 + bs + k)P(s)$$

The transfer function $G_1(s)$ between $X_1(s)$ and $P(s)$ can thus be obtained as

$$G_1(s) = \frac{X_1(s)}{P(s)} = \frac{m_2 s^2 + bs + k}{(m_1 s^2 + k)(m_2 s^2 + bs + k) + (bs + k)m_2 s^2}$$

Hence

$$G_1(j\omega) = \frac{-m_2\omega^2 + j\omega b + k}{(-m_1\omega^2 + k)(-m_2\omega^2 + j\omega b + k) + (bj\omega + k)m_2(-\omega^2)}$$

$$= \frac{(k - m_2\omega^2) + j\omega b}{(k - m_1\omega^2)(k - m_2\omega^2) - \omega^2 k m_2 + j\omega[bk - (m_1 + m_2)b\omega^2]}$$

from which we obtain

$$\left|G_1(j\omega)\right| = \frac{\sqrt{(k - m_2\omega^2)^2 + \omega^2 b^2}}{\sqrt{[(k - m_1\omega^2)(k - m_2\omega^2) - \omega^2 k m_2]^2 + \omega^2[bk - (m_1 + m_2)b\omega^2]^2}}$$

$$\angle G_1(j\omega) = \tan^{-1}\left(\frac{\omega b}{k - m_2\omega^2}\right) - \tan^{-1}\left\{\frac{\omega[bk - (m_1 + m_2)b\omega^2]}{(k - m_1\omega^2)(k - m_2\omega^2) - \omega^2 k m_2}\right\}$$

The steady-state output $x_1(t)$ can, therefore, be given by

$$x_1(t) = \left|G_1(j\omega)\right| P \sin\left[\omega t + \angle G_1(j\omega)\right]$$

Next, referring to Equation (3) we have the transfer function $G_2(s)$ between $X_2(s)$ and $P(s)$ as follows:

$$G_2(s) = \frac{X_2(s)}{P(s)}$$

$$= \frac{X_2(s)}{X_1(s)} \frac{X_1(s)}{P(s)}$$

$$= \frac{bs + k}{m_2 s^2 + bs + k} \frac{X_1(s)}{P(s)}$$

$$= \frac{bs + k}{(m_1 s^2 + k)(m_2 s^2 + bs + k) + (bs + k)m_2 s^2}$$

Hence

$$G_2(j\omega) = \frac{bj\omega + k}{(k - m_1\omega^2)(k - m_2\omega^2) - \omega^2 k m_2 + j\omega[bk - (m_1 + m_2)b\omega^2]}$$

The magnitude and angle of $G_2(j\omega)$ are given by

$$\left|G_2(j\omega)\right| = \frac{\sqrt{k^2 + b^2\omega^2}}{\sqrt{[(k - m_1\omega^2)(k - m_2\omega^2) - \omega^2 k m_2]^2 + \omega^2[bk - (m_1 + m_2)b\omega^2]^2}}$$

$$\angle G_2(j\omega) = \tan^{-1}\frac{b\omega}{k} - \tan^{-1}\left\{\frac{\omega[bk - (m_1 + m_2)b\omega^2]}{(k - m_1\omega^2)(k - m_2\omega^2) - \omega^2 k m_2}\right\}$$

The steady-state output $x_2(t)$ can be given by

$$x_2(t) = \left|G_2(j\omega)\right| P \sin\left[\omega t + \angle G_2(j\omega)\right]$$

B-9-5. tension $= m\omega^2 r = 0.1 \times 6.28^2 \times 1 = 3.94$ N

The tension in the cord is 3.94 N. The maximum angular speed can be obtained by solving the following equation for ω.

$$10 = 0.1 \times \omega^2 \times 1$$

The result is

$$\omega = \sqrt{100} = 10 \text{ rad/s} = 1.59 \text{ Hz}$$

The maximum angular speed that can be attained without breaking the cord is 1.59 Hz.

B-9-6. From the diagram shown below, we obtain

$$\frac{\text{centrifugal force}}{\text{gravitational force}} = \frac{m\omega^2 r}{mg} = \frac{0.15}{0.2598}$$

or

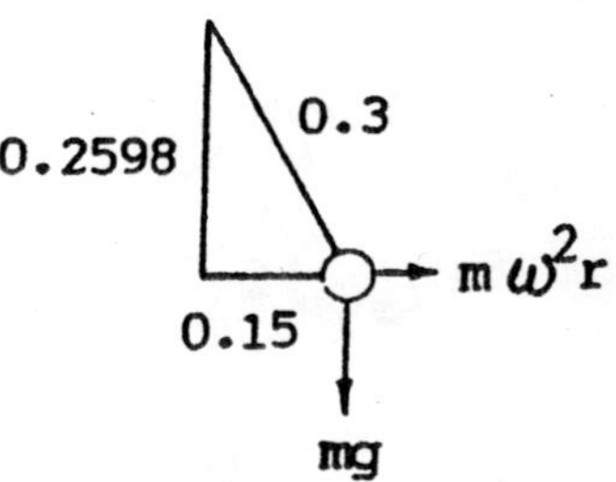

$$\frac{\omega^2 r}{g} = 0.5774$$

Solving for ω, we obtain

$$\omega = \sqrt{\frac{0.5774 \times 9.807}{0.15}} = 6.144 \text{ rad/s}$$

B-9-7. The equations of motion for the system are

$$M\ddot{x} + b\dot{x} + kx = m\omega^2 r \sin \omega t$$

$$f(t) = b\dot{x} + kx$$

If 10 % of the excitation force is to be transmitted to the foundation, the transmissibility must be equal to 0.1. Thus

$$TR = \frac{F_t}{F_o} = \frac{\sqrt{1 + (2\zeta\beta)^2}}{\sqrt{(1 - \beta^2)^2 + (2\zeta\beta)^2}} = 0.1$$

Since ζ is desired to be 0.2, we substitute $\zeta = 0.2$ into this last equation.

$$\frac{1 + (0.4\beta)^2}{(1 - \beta^2)^2 + (0.4\beta)^2} = 0.01$$

or

$$\beta^4 - 17.84\beta^2 - 99 = 0$$

Solving for β^2, we find

$$\beta^2 = 22.28 \qquad \text{or} \qquad -4.443$$

Noting that $\beta > 0$, we must have $\beta^2 = 22.28$. Then

$$\omega_n = \sqrt{\frac{k}{M}} = \sqrt{\frac{k}{100}}, \qquad \beta = \frac{\omega}{\omega_n} = \frac{62.8}{\sqrt{\frac{k}{100}}} = \sqrt{22.28} = 4.72$$

So we obtain

$$\sqrt{k} = \frac{628}{4.72} = 133$$

or

$$k = 17.7 \times 10^3 \text{ N/m}$$

The amplitude of force F_t transmitted to the foundation is

$$|F_t| = (m\omega^2 r)(TR) = 0.2 \times 62.8^2 \times 0.5 \times 0.1 = 39.44 \text{ N}$$

B-9-8. The equation of motion for the system is

$$m\ddot{x} + b(\dot{x} - \dot{y}) + k(x - y) = 0$$

Rewriting,

$$m(\ddot{x} - \ddot{y}) + b(\dot{x} - \dot{y}) + k(x - y) = - m\ddot{y}$$

By substituting $x - y = z$ into this last equation, we obtain

$$m\ddot{z} + b\dot{z} + kz = - m\ddot{y}$$

The Laplace transform of this equation, assuming zero initial conditions, is

$$(ms^2 + bs + k)Z(s) = - ms^2 Y(s)$$

or

$$\frac{Z(s)}{Y(S)} = \frac{- ms^2}{ms^2 + bs + k}$$

For the sinusoidal input $y = Y \sin \omega t$,

$$\frac{Z(j\omega)}{Y(j\omega)} = \frac{m\omega^2}{- m\omega^2 + bj\omega + k} = \frac{\omega^2}{-\omega^2 + 2\zeta\omega_n j\omega + \omega_n^2}$$

The steady-state amplitude ratio of z to y is

$$\left|\frac{Z(j\omega)}{Y(j\omega)}\right| = \frac{m\omega^2}{\sqrt{(k - m\omega^2)^2 + b^2\omega^2}} = \frac{\omega^2}{\sqrt{(\omega_n^2 - \omega^2)^2 + (2\zeta\omega_n\omega)^2}}$$

If $\omega \gg \omega_n$,

$$\left| \frac{Z(j\omega)}{Y(j\omega)} \right| = \frac{\omega^2}{\omega^2} = 1$$

Thus, the amplitude of sinusoidal displacement y of the base is equal to the amplitude of the relative displacement z.

If $\omega \ll \omega_n$, we have

$$\frac{Z(s)}{Y(s)} = -\frac{ms^2}{k}$$

or

$$\frac{Z(s)}{s^2 Y(s)} = -\frac{m}{k} = -\frac{1}{\omega_n^2}$$

So the acceleration $\ddot{y}$ of the base is proportional to z.

B-9-9. Define the displacement of spring k_2 as y. Then the equations of motion for the system are

$$m\ddot{x} + b_2(\dot{x} - \dot{y}) + k_1 x = p(t) \qquad (1)$$

$$b_2(\dot{x} - \dot{y}) = k_2 y \qquad (2)$$

The force f(t) transmitted to the foundation is

$$f(t) = k_1 x + k_2 y \qquad (3)$$

By taking Laplace transforms of Equations (1) and (2), assuming zero initial conditions, we obtain

$$(ms^2 + b_2 s + k_1)X(s) = b_2 sY(s) + P(s)$$

$$b_2 sX(s) = (b_2 s + k_2)Y(s)$$

By eliminating Y(s) from the last two equations and simplifying, we get

$$\frac{X(s)}{P(s)} = \frac{b_2 s + k_2}{(ms^2 + k_1)(b_2 s + k_2) + b_2 k_2 s}$$

The Laplace transform of Equation (3) is

$$F(s) = k_1 X(s) + k_2 Y(s) = k_1 X(s) + \frac{b_2 k_2 s}{b_2 s + k_2} X(s)$$

$$= \frac{k_1 b_2 s + k_1 k_2 + b_2 k_2 s}{b_2 s + k_2} X(s)$$

So we have

$$\frac{F(s)}{X(s)} = \frac{(k_1 b_2 + k_2 b_2)s + k_1 k_2}{b_2 s + k_2}$$

and

$$\frac{F(s)}{P(s)} = \frac{F(s)}{X(s)} \frac{X(s)}{P(s)} = \frac{(k_1 + k_2)b_2 s + k_1 k_2}{(ms^2 + k_1)(b_2 s + k_2) + b_2 k_2 s}$$

The force transmissibility TR is

$$TR = \left| \frac{(k_1 + k_2)b_2 j\omega + k_1 k_2}{(-m\omega^2 + k_1)(k_2 + b_2 j\omega) + b_2 k_2 j\omega} \right|$$

$$= \frac{\sqrt{(k_1 + k_2)^2 b_2^2 \omega^2 + k_1^2 k_2^2}}{\sqrt{(k_1 - m\omega^2)^2 k_2^2 + [b_2 k_2 \omega + (k_1 - m\omega^2) b_2 \omega]^2}}$$

The amplitude of the force transmitted to the foundation is

$$\left| F(j\omega) \right| = \frac{P\sqrt{(k_1 + k_2)^2 b_2^2 \omega^2 + k_1^2 k_2^2}}{\sqrt{(k_1 - m\omega^2)^2 k_2^2 + [b_2 k_2 \omega + (k_1 - m\omega^2) b_2 \omega]^2}}$$

B-9-10. Define the displacement of the top end of spring k_2 as z. Then the equations of motion for the system are

$$m\ddot{x} + b_2(\dot{x} - \dot{z}) + k_1(x - p) = 0$$

$$b_2(\dot{x} - \dot{z}) = k_2(z - p)$$

where $p = P \sin \omega t$. Rewriting these equations,

$$m\ddot{x} + b_2\dot{x} + k_1 x = b_2\dot{z} + k_1 p$$

$$b_2\dot{x} + k_2 p = k_2 z + b_2 \dot{z}$$

Laplace transforming these two equations, assuming zero initial conditions, we obtain

$$(ms^2 + b_2 s + k_1)X(s) = b_2 sZ(s) + k_1 P(s)$$

$$b_2 sX(s) + k_2 P(s) = (k_2 + b_2 s)Z(s)$$

Eliminating Z(s) from the last two equations and simplifying gives

$$[(ms^2 + k_1)(k_2 + b_2 s) + k_2 b_2 s]X(s) = b_2 k_2 sP(s) + k_1 P(s)(k_2 + b_2 s)$$

So the transfer function X(s)/P(s) is obtained as

$$\frac{X(s)}{P(s)} = \frac{b_2k_2s + k_1k_2 + k_1b_2s}{(ms^2 + k_1)(b_2s + k_2) + b_2k_2s}$$

The motion transmissibility TR is

$$TR = \left|\frac{X(j\omega)}{P(j\omega)}\right| = \frac{\sqrt{(k_1 + k_2)^2b_2^2\omega^2 + k_1^2k_2^2}}{\sqrt{k_2^2(k_1 - m\omega^2)^2 + [b_2k_2\omega + b_2\omega(k_1 - m\omega^2)]^2}}$$

The vibration amplitude $|X(j\omega)|$ of the machine is

$$|X(j\omega)| = P\cdot TR$$

$$= \frac{P\sqrt{(k_1 + k_2)^2b_2^2\omega^2 + k_1^2k_2^2}}{\sqrt{k_2^2(k_1 - m\omega^2)^2 + [b_2k_2\omega + b_2\omega(k_1 - m\omega^2)]^2}}$$

B-9-11. The equations of motion for the system are

$$m\ddot{x} + b\dot{x} + kx + k_a(x - y) = p = P \sin\omega t$$

$$m_a\ddot{y} + k_a(y - x) = 0$$

Laplace transforming these two equations, assuming zero initial conditions, we obtain

$$(ms^2 + bs + k + k_a)X(s) = k_aY(s) + P(s)$$

$$(m_as^2 + k_a)Y(s) = k_aX(s)$$

Eliminating Y(s) from the last two equations and simplifying, we obtain

$$\frac{X(s)}{P(s)} = \frac{m_as^2 + k_a}{(ms^2 + bs + k)(m_as^2 + k_a) + m_ak_as^2}$$

Hence

$$\frac{X(j\omega)}{P(j\omega)} = \frac{k_a - m_a\omega^2}{(k - m\omega^2 + bj\omega)(k_a - m_a\omega^2) - m_ak_a\omega^2}$$

Note that if $\sqrt{k_a/m_a} = \omega$, then $X(j\omega) = 0$. Since

$$\frac{Y(s)}{X(s)} = \frac{k_a}{m_a s^2 + k_a}$$

we have

$$\frac{Y(j\omega)}{P(j\omega)} = \frac{k_a}{(k - m\omega^2 + bj\omega)(k_a - m_a\omega^2) - m_a k_a \omega^2}$$

By substituting $k_a/m_a = \omega^2$ into this last equation, we get

$$\frac{Y(j\omega)}{P(j\omega)} = \frac{k_a}{-m_a k_a \omega^2} = -\frac{1}{m_a \omega^2}$$

Hence

$$|Y(j\omega)| = \frac{|P(j\omega)|}{m_a \omega^2} = \frac{P}{m_a \frac{k_a}{m_a}} = \frac{P}{k_a}$$

The amplitude of vibration of mass m_a is P/k_a.

B-9-12. Assuming small angles θ_1 and θ_2 the equations of motion for the system may be obtained as follows:

$$m_1 \ell^2 \ddot{\theta}_1 = -m_1 g \ell \theta_1 - ka^2(\theta_1 - \theta_2)$$

$$m_2 \ell^2 \ddot{\theta}_2 = -m_2 g \ell \theta_2 - ka^2(\theta_2 - \theta_1)$$

Rewriting these equations, we obtain

$$m_1 \ell^2 \ddot{\theta}_1 + m_1 g \ell \theta_1 + ka^2\theta_1 = ka^2\theta_2$$

$$m_2 \ell^2 \ddot{\theta}_2 + m_2 g \ell \theta_2 + ka^2\theta_2 = ka^2\theta_1$$

which can be simplified to

$$\ddot{\theta}_1 + \left(\frac{g}{\ell} + \frac{ka^2}{m_1\ell^2}\right)\theta_1 = \frac{ka^2}{m_1\ell^2}\theta_2 \qquad (1)$$

$$\ddot{\theta}_2 + \left(\frac{g}{\ell} + \frac{ka^2}{m_2\ell^2}\right)\theta_2 = \frac{ka^2}{m_2\ell^2}\theta_1 \qquad (2)$$

To find the natural frequencies of the free vibration, we assume the motion to be harmonic. That is, we assume

$$\theta_1 = A \sin \omega t, \qquad \theta_2 = B \sin \omega t$$

Then

$$\ddot{\theta}_1 = -A\omega^2 \sin \omega t, \qquad \ddot{\theta}_2 = -B\omega^2 \sin \omega t$$

Substituting the preceding expressions into Equations (1) and (2), we obtain

$$\left[- A\omega^2 + \left(\frac{g}{\ell} + \frac{ka2}{m_1 \ell^2} \right) A - \frac{ka^2}{m_1 \ell^2} B \right] \sin \omega t = 0$$

$$\left[- B\omega^2 + \left(\frac{g}{\ell} + \frac{ka^2}{m_2 \ell^2} \right) B - \frac{ka^2}{m_2 \ell^2} A \right] \sin \omega t = 0$$

Since these equations must be satisfied at all times and $\sin \omega t$ cannot be zero at all times, the quantities in the brackets must be equal to zero. Thus

$$\left(\frac{g}{\ell} + \frac{ka^2}{m_1 \ell^2} - \omega^2 \right) A - \frac{ka^2}{m_1 \ell^2} B = 0 \tag{3}$$

$$- \frac{ka^2}{m_2 \ell^2} A + \left(\frac{g}{\ell} + \frac{ka^2}{m_2 \ell^2} - \omega^2 \right) B = 0 \tag{4}$$

For constants A and B to be nonzero, the determinant of the coefficients of Equations (3) and (4) must be equal to zero, or

$$\begin{vmatrix} \frac{g}{\ell} + \frac{ka^2}{m_1 \ell^2} - \omega^2 & - \frac{ka^2}{m_1 \ell^2} \\ - \frac{ka^2}{m_2 \ell^2} & \frac{g}{\ell} + \frac{ka^2}{m_2 \ell^2} - \omega^2 \end{vmatrix} = 0 \tag{5}$$

This determinant equation determines the natural frequencies of the system. Equation (5) can be rewritten as

$$\left(\frac{g}{\ell} + \frac{ka^2}{m_1 \ell^2} - \omega^2 \right) \left(\frac{g}{\ell} + \frac{ka^2}{m_2 \ell^2} - \omega^2 \right) - \frac{ka^2}{m_1 \ell^2} \frac{ka^2}{m_2 \ell^2} = 0$$

or

$$\omega^4 + \left(- \frac{2g}{\ell} - \frac{ka^2}{m_1 \ell^2} - \frac{ka^2}{m_2 \ell^2} \right) \omega^2 + \left(\frac{g}{\ell} \right)^2 + \left(\frac{ka^2}{m_1 \ell^2} + \frac{ka^2}{m_2 \ell^2} \right) \frac{g}{\ell} = 0$$

This last equation can be factored as follows:

$$\left(\omega^2 - \frac{g}{\ell} - \frac{ka^2}{m_1\ell^2} - \frac{ka^2}{m_2\ell^2}\right)\left(\omega^2 - \frac{g}{\ell}\right) = 0$$

or

$$\omega^2 = \frac{g}{\ell}, \qquad \omega^2 = \frac{g}{\ell} + \frac{ka^2}{m_1\ell^2} + \frac{ka^2}{m_2\ell^2}$$

Thus

$$\omega_1 = \sqrt{\frac{g}{\ell}}, \qquad \omega_2 = \sqrt{\frac{g}{\ell} + \frac{ka^2}{m_1\ell^2} + \frac{ka^2}{m_2\ell^2}}$$

The first natural frequency is ω_1 (first mode) and the second natural frequency is ω_2 (second mode).

At the first natural frequency $\omega = \omega_1 = \sqrt{g/\ell}$, we obtain from Equation (3) the following expression:

$$\frac{A}{B} = \frac{\dfrac{ka^2}{m_1\ell^2}}{\dfrac{g}{\ell} + \dfrac{ka^2}{m_1\ell^2} - \dfrac{g}{\ell}} = 1$$

Note that we obtain the same result from Equation (4). Thus, at the first mode the amplitude ratio A/B becomes unity, or A = B. This means thatboth masses move the same amount in the same direction. This mode is depicted in Figure (a) below.

At the second natural frequency $\omega = \omega_2$ we obtain from Equation (3)

$$\frac{A}{B} = \frac{\dfrac{ka^2}{m_1\ell^2}}{\dfrac{g}{\ell} + \dfrac{ka^2}{m_1\ell^2} - \dfrac{g}{\ell} - \dfrac{ka^2}{m_1\ell^2} - \dfrac{ka^2}{m_2\ell^2}} = -\frac{m_2}{m_1}$$

[We obtain the same result from Equation (4).] At the second mode, the amplitude ratio A/B becomes $-m_2/m_1$ or $A = -(m_2/m_1)B$. This means that masses move in the opposite direction. This mode is depicted in Figure (b) shown below.

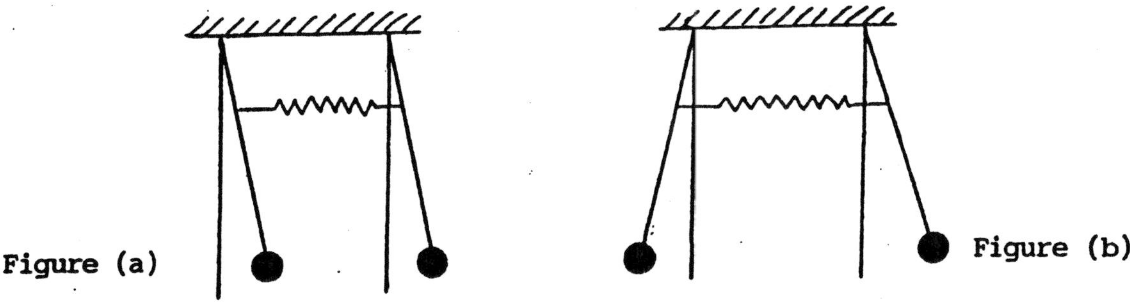

B-9-13. The equations of motion for the system are

$$m\ddot{x} = -(x + \ell_1 \sin\theta)k_1 - (x - \ell_2 \sin\theta)k_2$$

$$J\ddot{\theta} = -(x + \ell_1 \sin\theta)k_1 \ell_1 \cos\theta + (x - \ell_2 \sin\theta)k_2 \ell_2 \cos\theta$$

For small angle θ, we have $\sin\theta \doteqdot \theta$ and $\cos\theta \doteqdot 1$ and the preceding two equations become

$$m\ddot{x} + (k_1 + k_2)x + (\ell_1 k_1 - \ell_2 k_2)\theta = 0 \qquad (1)$$

$$J\ddot{\theta} + (\ell_1^2 k_1 + \ell_2^2 k_2)\theta + (\ell_1 k_1 - \ell_2 k_2)x = 0 \qquad (2)$$

Notice that if $\ell_1 k_1 = \ell_2 k_2$, then the coupling terms become zero and two equations become independent. However, in this problem

$$\ell_1 = 1.5 \text{ m}, \qquad \ell_2 = 2 \text{ m}, \qquad k_1 = k_2 = 4 \times 10^4 \text{ N/m}$$

Thus,

$$\ell_1 k_1 - \ell_2 k_2 = -0.5 \times 4 \times 10^4 \neq 0$$

Therefore, coupling exists between Equation (1) and Equation (2).

To find the natural frequencies for the system, assume the following harmonic motion:

$$x = A \sin\omega t, \qquad \theta = B \sin\omega t$$

Then, from Equations (1) and (2) we obtain

$$(k_1 + k_2 - m\omega^2)A + (\ell_1 k_1 - \ell_2 k_2)B = 0 \qquad (3)$$

$$(\ell_1 k_1 - \ell_2 k_2)A + (\ell_1^2 k_1 + \ell_2^2 k_2 - J\omega^2)B = 0 \qquad (4)$$

For amplitudes A and B to be nonzero, the determinant of the coefficients of Equations (3) and (4) must be equal to zero, or

$$\begin{vmatrix} k_1 + k_2 - m\omega^2 & \ell_1 k_1 - \ell_2 k_2 \\ \ell_1 k_1 - \ell_2 k_2 & \ell_1^2 k_1 + \ell_2^2 k_2 - J\omega^2 \end{vmatrix} = 0 \qquad (5)$$

This determinant equation determines the natural frequencies of the system. Equation (5) can be rewritten as

$$(k_1 + k_2 - m\omega^2)(\ell_1^2 k_1 + \ell_2^2 k_2 - J\omega^2) - (\ell_1 k_1 - \ell_2 k_2)^2 = 0$$

or

$$mJ\omega^4 - [(k_1 + k_2)J + (\ell_1^2 k_1 + \ell_2^2 k_2)m]\omega^2 + k_1 k_2(\ell_1 + \ell_2)^2 = 0$$

which can be simplified to

$$\omega^4 - \left(\frac{k_1 + k_2}{m} + \frac{\ell_1^2 k_1 + \ell_2^2 k_2}{J}\right)\omega^2 + \frac{k_1 k_2(\ell_1 + \ell_2)^2}{mJ} = 0 \qquad (6)$$

Notice that this last equation determines the natural frequencies of the system. By substituting the given numerical values for $\ell_1, \ell_2, k_1, k_2, m$, and J into Equation (6), we obtain

$$\omega^4 - 140\,\omega^2 + 3920 = 0$$

Solving this equation for ω^2, we get

$$\omega^2 = 38.695 \qquad \text{or} \qquad 101.305$$

Hence

$$\omega_1 = 6.2205, \qquad \omega_2 = 10.065$$

The first natural frequency is $\omega_1 = 6.2205$ rad/s and the second natural frequency is $\omega_2 = 10.065$ rad/s.

To determine the modes of vibration, notice that from Equations (3) and (4) we have

$$\frac{A}{B} = \frac{\ell_2 k_2 - \ell_1 k_1}{k_1 + k_2 - m\omega^2} = \frac{\ell_1^2 k_1 + \ell_2^2 k_2 - J\omega^2}{\ell_2 k_2 - \ell_1 k_1}$$

By substituting the given numerical values into this last equation, we obtain

$$\frac{A}{B} = \frac{2}{8 - 0.2\,\omega^2} = \frac{25 - 0.25\,\omega^2}{2} \qquad (7)$$

For the first mode of vibration ($\omega_1 = 6.2205$ rad/s) the amplitude ratio A/B becomes as follows:

$$\frac{A}{B} = \frac{2}{8 - 0.2 \times 6.2205^2} = \frac{25 - 0.25 \times 6.2205^2}{2}$$

$$= \frac{2}{0.261} = \frac{15.326}{2} = 7.663$$

Notice that the ratio of the displacements of springs k_1 and k_2 are

$$\frac{x + \ell_1\theta}{x - \ell_2\theta} = \frac{A + \ell_1 B}{A - \ell_2 B} = \frac{(7.663 + 1.5)B}{(7.663 - 2)B} = \frac{9.163B}{5.663B}$$

The first mode of vibration is shown in Figure (a) on next page.

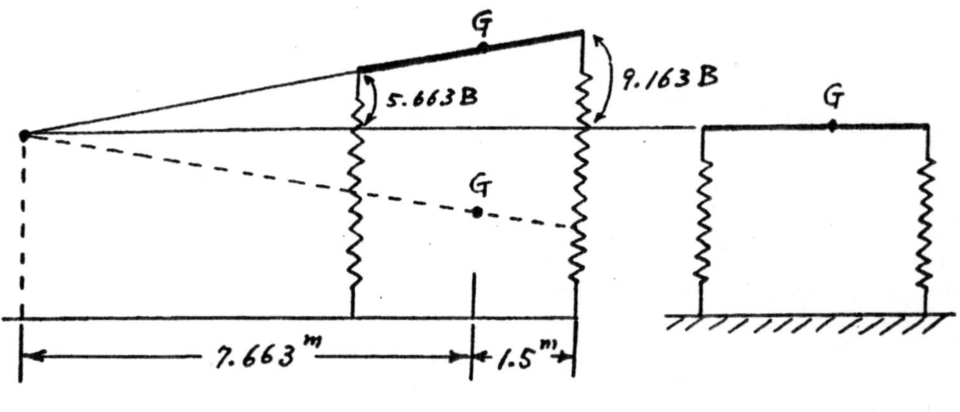

Figure (a)

For the second mode of vibration (ω_2 = 10.065 rad/s) the amplitude ratio A/B becomes as

$$\frac{A}{B} = \frac{2}{8 - 0.2 \times 10.065^2} = \frac{25 - 0.25 \times 10.065^2}{2}$$

$$= \frac{2}{-12.261} = \frac{-0.32625}{2} = -\ 0.1631$$

Hence the ratio of the displacements of springs k_1 and k_2 becomes

$$\frac{x + \ell_1\theta}{x - \ell_2\theta} = \frac{A + \ell_1 B}{A - \ell_2 B} = \frac{(-0.1631 + 1.5)B}{(-0.1631 - 2)B} = \frac{1.3369B}{-2.1631B}$$

The second mode of vibration is shown in Figure (b) below.

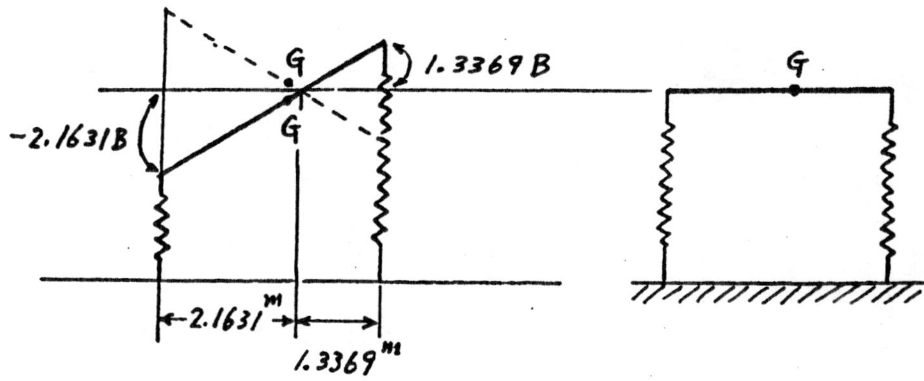

Figure (b)

B-9-14. The system shown in Figure 9-50 is a special case of the system shown in Figure 9-32 (Problem A-9-14). By defining

$$k_1 = k, \quad k_2 = 2k, \quad k_3 = k, \quad m_1 = m, \quad m_2 = m$$

Equations (9-40) and (9-41) become as follows:

$$\frac{A}{B} = \frac{2k}{-m\omega^2 + k + 2k} = \frac{2k}{-m\omega^2 + 3k} \tag{1}$$

$$\frac{A}{B} = \frac{-m\omega^2 + 2k + k}{2k} = \frac{-m\omega^2 + 3k}{2k} \tag{2}$$

Also, ω^2 that satisfies Equations (9-38) and (9-39) becomes as follows:

$$\omega^2 = \frac{1}{2}\left(\frac{k + 2k}{m} + \frac{2k + k}{m}\right) \pm \sqrt{\frac{1}{4}\left(\frac{k + 2k}{m} - \frac{2k + k}{m}\right)^2 + \frac{4k^2}{m^2}}$$

$$= \frac{3k}{m} \pm \frac{2k}{m} = \frac{k}{m} \text{ or } \frac{5k}{m}$$

Define

$$\omega_1^2 = \frac{k}{m}, \qquad \omega_2^2 = \frac{5k}{m}$$

By substituting ω_1^2 into Equation (1), we obtain

$$\frac{A_1}{B_1} = \frac{2k}{-m(k/m) + 3k} = \frac{2k}{2k} = 1$$

Similarly, by substituting ω_2^2 into Equation (1), we get

$$\frac{A_2}{B_2} = \frac{2k}{-m(5k/m) + 3k} = \frac{2k}{-2k} = -1$$

[We get the same result if we substitute ω_1^2 or ω_2^2 into Equation (2).] Hence we have

$$\frac{A_1}{B_1} = 1 > 0, \qquad \frac{A_2}{B_2} = -1 < 0$$

which means that in the first mode of vibration (with frequency ω_1), the masses move in the same direction by the same amount. In the second mode of vibration (with frequency ω_2), the masses move in opposite directions by the same amount. Figures (a) and (b) shown on next page depict the first mode of vibration and second mode of vibration, respectively.

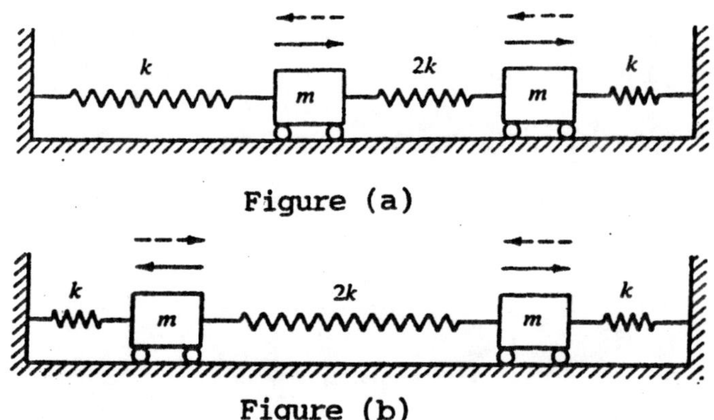

Figure (a)

Figure (b)

B-9-15. The equations of motion for the system are

$$M\ddot{x} + k_1(x - y) + k_2 x = 0$$

$$m\ddot{y} + k_1(y - x) = 0$$

Substituting the given numerical values into these two equations and simplifying, we have

$$\ddot{x} + 11x - y = 0 \qquad (1)$$

$$\ddot{y} + 10y - 10x = 0 \qquad (2)$$

To find the natural frequencies of the free vibration, assume that the motion is harmonic, or

$$x = A \sin \omega t, \qquad y = B \sin \omega t$$

Then

$$\ddot{x} = -A\omega^2 \sin \omega t, \qquad \ddot{y} = -B\omega^2 \sin \omega t$$

If the preceeding expressions are substituted into Equations (1) and (2), we obtain

$$(-A\omega^2 + 11A - B) \sin \omega t = 0$$

$$(-B\omega^2 + 10B - 10A) \sin \omega t = 0$$

Since these two equations must be satisfied at all times and $\sin \omega t$ cannot be zero at all times, we must have

$$-A\omega^2 + 11A - B = 0$$

$$-B\omega^2 + 10B - 10A = 0$$

Rearranging,

$$(11 - \omega^2)A - B = 0 \qquad (3)$$

$$-10A + (10 - \omega^2)B = 0 \qquad (4)$$

For constants A and B to be nonzero, the determinant of the coefficient matrix must be equal to zero, or

$$\begin{vmatrix} 11 - \omega^2 & -1 \\ -10 & 10 - \omega^2 \end{vmatrix} = 0$$

which yields

$$(11 - \omega^2)(10 - \omega^2) - 10 = 0$$

or

$$\omega^4 - 21\omega^2 + 100 = 0$$

which can be rewritten as

$$(\omega^2 - 7.2985)(\omega^2 - 13.7016) = 0$$

Hence,

$$\omega_1^2 = 7.2985, \qquad \omega_2^2 = 13.7016$$

or

$$\omega_1 = 2.7016, \qquad \omega_2 = 3.7016$$

The frequency of the first mode is 2.7016 rad/s and the frequency of the second mode is 3.7016 rad/s.

From Equations (3) and (4), we obtain

$$\frac{A}{B} = \frac{1}{11 - \omega^2}, \qquad \frac{A}{B} = \frac{10 - \omega^2}{10}$$

By substituting $\omega_1^2 = 7.2985$ into A/B, we obtain

$$\frac{A}{B} = \frac{1}{11 - 7.2985} = \frac{10 - 7.2985}{10} = 0.27016 > 0$$

Similarly, by substituting $\omega_2^2 = 13.7016$ into A/B, we get

$$\frac{A}{B} = \frac{1}{11 - 13.7016} = \frac{10 - 13.7016}{10} = -0.37016 < 0$$

Hence, at the first mode of vibration, two masses move in the same direction, while at the second mode of vibration, two masses move in opposite directions.

Next, we shall obtain the vibrations x(t) and y(t) subjected to the given initial conditions. Laplace transforming Equations (1) and (2),

$$[s^2X(s) - sx(0) - \dot{x}(0)] + 11X(s) - Y(s) = 0$$

$$[s^2Y(s) - sy(0) - \dot{y}(0)] + 10Y(s) - 10X(s) = 0$$

Substituting the given initial conditions into the preceeding two equations we get

$$(s^2 + 11)X(s) = 0.05s + Y(s) \quad (5)$$

$$(s^2 + 10)Y(s) = 10X(s) \quad (6)$$

Eliminating Y(s) from Equations (5) and (6),

$$(s^2 + 11)X(s) = 0.05s + \frac{10}{s^2 + 10} X(s)$$

which can be simplified to

$$X(s) = \frac{0.05s(s^2 + 10)}{s^4 + 21s^2 + 100}$$

Similarly, we can obtain Y(s) as follows:

$$Y(s) = \frac{0.5s}{s^4 + 21s^2 + 100}$$

To obtain the responses x(t) and y(t) to the given initial conditions, we rewrite X(s) and Y(s) as follows:

$$X(s) = \frac{0.05s^4 + 0.5s^2}{s^4 + 21s^2 + 100} \frac{1}{s}, \quad Y(s) = \frac{0.5s^2}{s^4 + 21s^2 + 100} \frac{1}{s}$$

Possible MATLAB programs to plot x(t) and y(t), respectively, are given next. The resulting plots x(t) versus t and y(t) versus t are shown on next page.

```
>> num1 = [0.05   0   0.5   0   0];
>> num2 = [0   0   0.5   0   0];
>> den = [1   0   21   0   100];
>> t = 0:0.05:15;
>> x = step(num1,den,t);
>> y = step(num2,den,t);
>> subplot(221); plot(t,x)
>> grid
>> title('Vibration x(t) due to initial conditions')
>> xlabel('t (sec)')
>> ylabel('x(t)')
>> subplot(222); plot(t,y)
>> grid
>> title('Vibration y(t) due to initial conditions')
>> xlabel('t (sec)')
>> ylabel('y(t)')
```

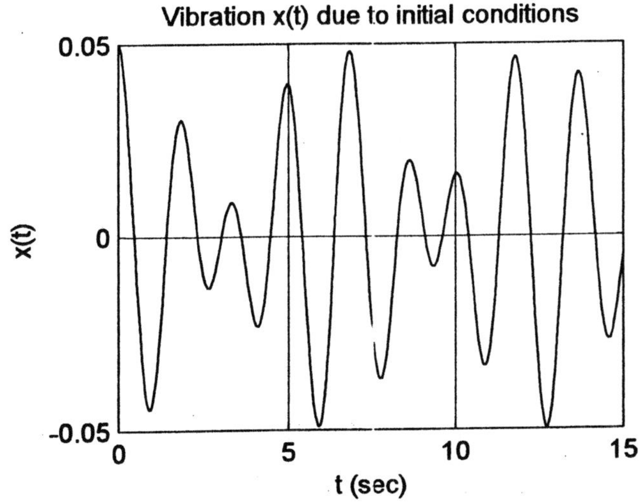

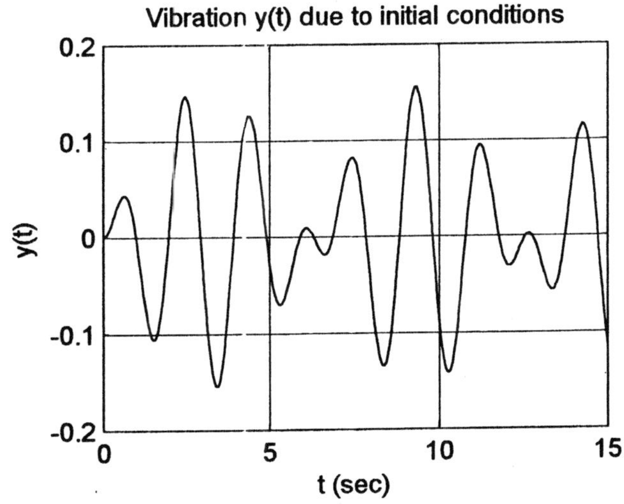

B-9-16. All necessary derivations of equations for the system are given in Problem A-9-16. The equations for the system are

$$(2s^2 + 50)X(s) = 2sx(0) + 10Y(s) \tag{1}$$

$$(s^2 + 10)Y(s) = sy(0) + 10X(s) \tag{2}$$

Referring to Equation (9-54) we have

$$X(s) = \frac{(s^2 + 10)sx(0) + 5sy(0)}{s^4 + 35s^2 + 200} \tag{3}$$

Case (a): For the initial conditions

$$x(0) = 0.028078, \quad \dot{x}(0) = 0, \quad y(0) = 0.1, \quad \dot{y}(0) = 0$$

Equation (3) becomes as follows:

$$X(s) = \frac{(s^2 + 10) \text{ x } 0.028078s + 0.5s}{s^4 + 35s^2 + 200}$$

$$= \frac{0.02807s(s^2 + 27.808)}{(s^2 + 27.808)(s^2 + 7.1922)}$$

$$= \frac{0.02807s}{s^2 + 7.1922}$$

$$= \frac{0.02807s^2}{s^2 + 7.1922} \frac{1}{s} \tag{4}$$

By substituting Equation (4) into Equation (2) and solving for Y(s), we obtain

$$Y(s) = \frac{1}{s^2 + 10}\left[sy(0) + \frac{0.28078s}{s^2 + 7.1922} \right]$$

Substituting $y(0) = 0.1$ into the last equation and simplifying, we get

$$Y(s) = \frac{0.1}{s^2 + 10} \frac{s(s^2 + 10)}{s^2 + 7.1922}$$

$$= \frac{0.1s}{s^2 + 7.1922}$$

$$= \frac{0.1s^2}{s^2 + 7.1922} \frac{1}{s}$$

To obtain plots of $x(t)$ versus t and $y(t)$ versus t, we may enter the following MATLAB program into the computer. The resulting plots are shown in Figure (a).

```
>> num1 = [0.02807   0    0];
>> num2 = [0.1    0    0];
>> den = [1    0    7.1922];
>> t = 0:0.01:8;
>> x = step(num1,den,t);
>> y = step(num2,den,t);
>> plot(t,x,t,y);
>> grid
>> title('Responses x(t) and y(t) due to initial conditions (a)')
>> xlabel('t (sec)')
>> ylabel('x(t) and y(t)')
>> text(2,-0.04,'x(t)')
>> text(3,0.075,'y(t)')
```

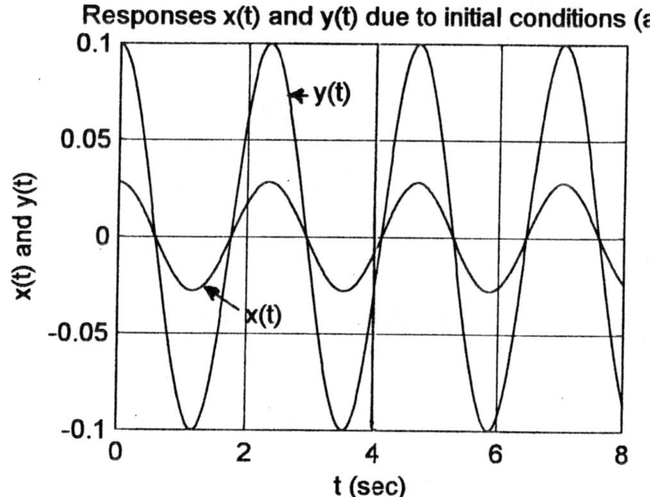

Figure (a)

Case (b): For the initial conditions

$$x(0) = 0.17808, \quad \dot{x}(0) = 0, \quad y(0) = -0.1, \quad \dot{y}(0) = 0$$

we obtain the following expressions for X(s) and Y(s):

$$X(s) = \frac{0.17808s}{s^2 + 27.808} = \frac{0.17808s^2}{s^2 + 27.808}\frac{1}{s}$$

$$Y(s) = -\frac{0.1s}{s^2 + 27.808} = -\frac{0.1s^2}{s^2 + 27.808}\frac{1}{s}$$

A MATLAB program for obtaining plots of x(t) versus t and y(t) versus t is shown below. The resulting plots are shown in Figure (b) below.

```
>> num1 = [0.17808   0   0];
>> num2 = [-0.1   0   0];
>> den = [1   0   27.808];
>> t = 0:0.01:4;
>> x = step(num1,den,t);
>> y = step(num2,den,t);
>> plot(t,x,t,y);
>> grid
>> title('Responses x(t) and y(t) due to initial conditions (b)')
>> xlabel('t (sec)')
>> ylabel('x(t) and y(t)')
>> text(1.5,0.15,'x(t)')
>> text(1.2,-0.15,'y(t)')
```

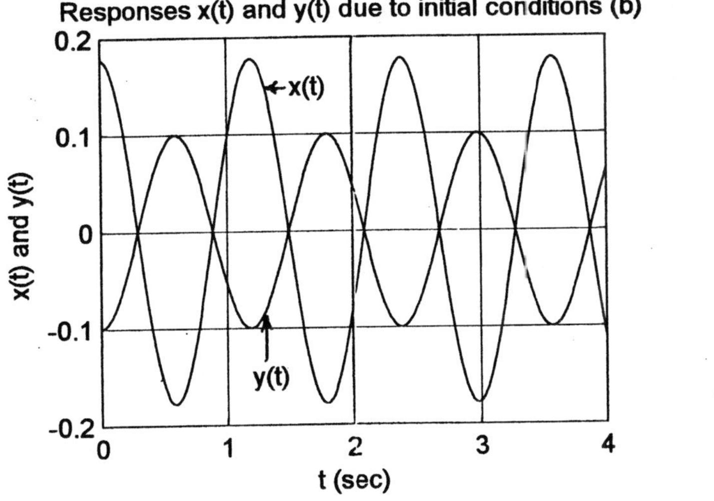

Figure (b)

Case (c): For the initial conditions

$$x(0) = 0.1, \quad \dot{x}(0) = 0, \quad y(0) = -0.1, \quad \dot{y}(0) = 0$$

we obtain the following expressions for X(s) and Y(s):

$$X(s) = \frac{(s^2 + 10)sx(0) + 5sy(0)}{s^4 + 35s^2 + 200}$$

$$= \frac{(s^2 + 10)s(0.1) + 5s(-0.1)}{s^4 + 35s^2 + 200}$$

$$= \frac{0.1s^3 + 0.5s}{s^4 + 35s^2 + 200}$$

$$= \frac{0.1s^4 + 0.5s^2}{s^4 + 35s^2 + 200} \frac{1}{s}$$

$$Y(s) = \frac{sy(0)}{s^2 + 10} + \frac{10X(s)}{s^2 + 10}$$

$$= -\frac{0.1}{s^2 + 10} \frac{s^5 + 25s^3 + 150s}{s^4 + 35s^2 + 200}$$

$$= -\frac{0.1s(s^2 + 15)}{s^4 + 35s^2 + 200}$$

$$= -\frac{0.1s^4 + 1.5s^2}{s^4 + 35s^2 + 200} \frac{1}{s}$$

A MATLAB program to obtain plots of x(t) versus t and y(t) versus t is shown below. The resulting plots are shown in Figure (c).

```
>> num1 = [0.1   0   0.5   0   0];
>> num2 = [-0.1   0   -1.5   0   0];
>> den = [1   0   35   0   200];
>> t = 0:0.01:6;
>> x = step(num1,den,t);
>> y = step(num2,den,t);
>> plot(t,x,t,y);
>> grid
>> title('Responses x(t) and y(t) due to initial conditions (c)')
>> xlabel('t (sec)')
>> ylabel('x(t) and y(t)')
>> text(0.3,0.12,'x(t)')
>> text(2.2,-0.16,'y(t)')
```

Responses x(t) and y(t) due to initial conditions (c)
x(t)
y(t)
x(t) and y(t)
t (sec)
0.15
0.1
0.05
0
-0.05
-0.1
-0.15
-0.2
0
2
4
6

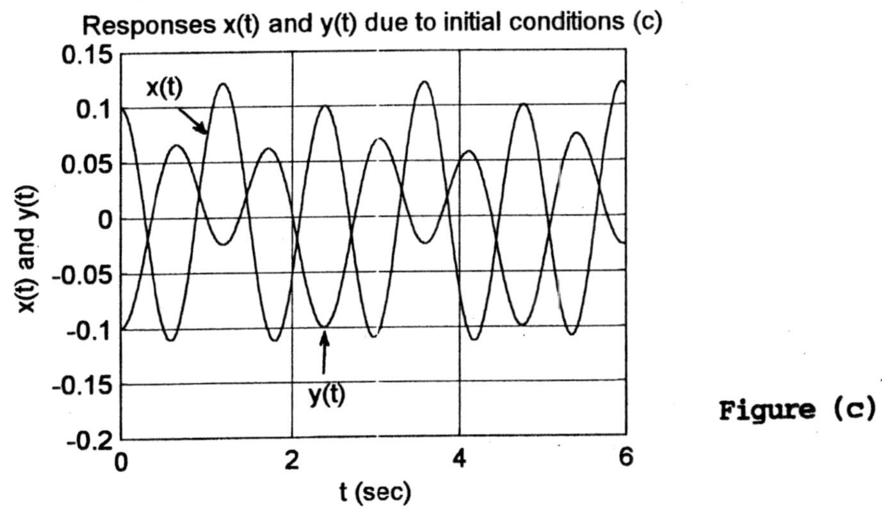

Figure (c)

CHAPTER 10

B-10-1. A simplified block diagram of the system is shown to the right. The transfer function C(s)/R(s) is

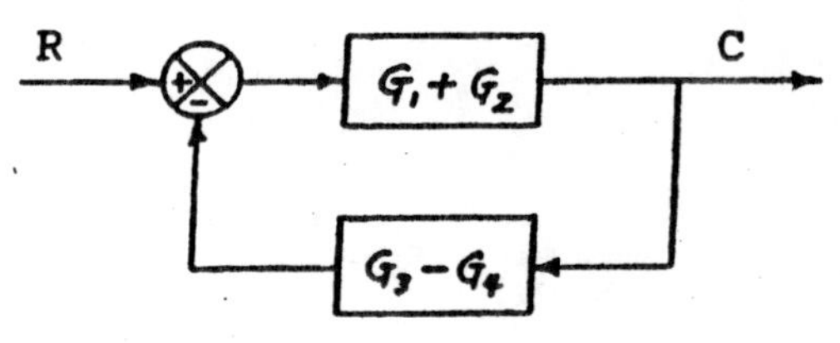

$$\frac{C(s)}{R(s)} = \frac{G_1 + G_2}{1 + (G_1 + G_2)(G_3 - G_4)}$$

B-10-2. Simplified block diagrams for the system are shown below.

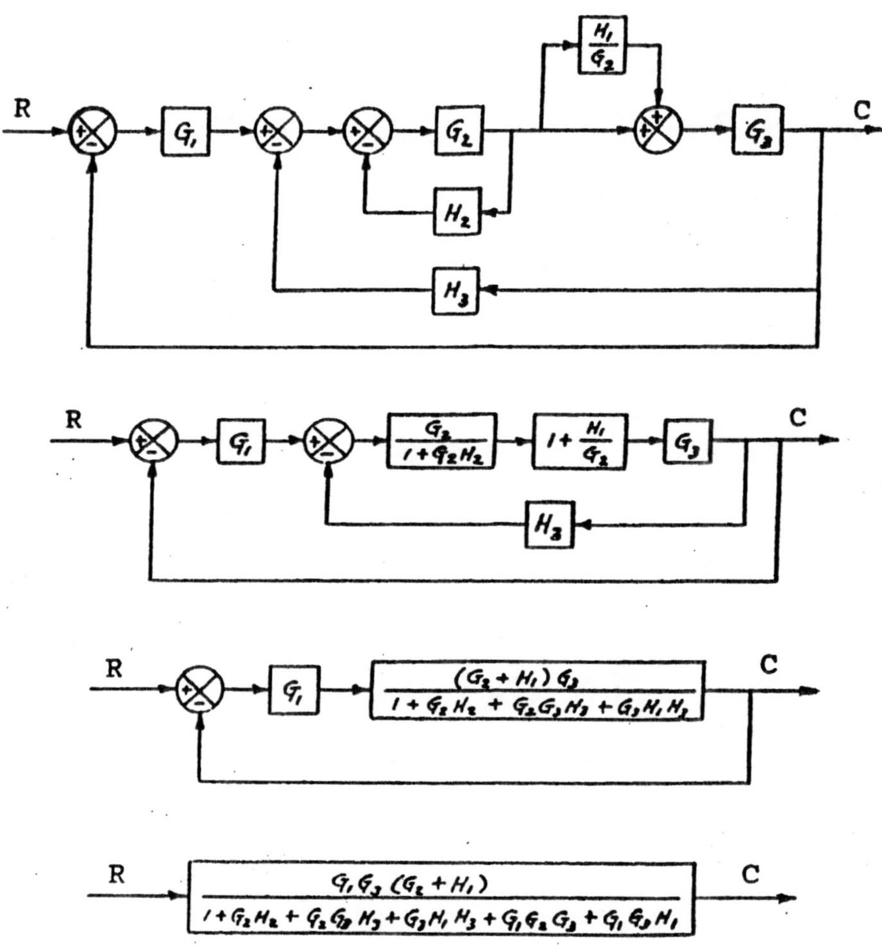

The transfer function C(s)/R(s) is

$$\frac{C(s)}{R(s)} = \frac{G_1G_3(G_2 + H_1)}{1 + G_2H_2 + G_2G_3H_3 + G_3H_1H_3 + G_1G_2G_3 + G_1G_3H_1}$$

B-10-3. A MATLAB program to solve this problem is given below.

```
>> num1 = [0.5   1];
>> den1 = 1;
>> sys1 = tf(num1,den1);
>> num2 = [10];
>> den2 = [1   2    0];
>> sys2 = tf(num2,den2);
>> sysg = series(sys1,sys2);
>> sys = feedback(sysg,[1])

Transfer function:
   5 s + 10
--------------
s^2 + 7 s + 10
```

The closed-loop transfer function obtained is

$$\frac{C(s)}{R(s)} = \frac{5s + 10}{s^2 + 7s + 10}$$

B-10-4. A MATLAB program to solve this problem is shown to the right.

The closed-loop transfer function obtained is

$$\frac{C(s)}{R(s)} = \frac{10s + 50}{s^3 + 8s^2 + 37s + 50}$$

```
>> num1 = [1   5];
>> den1 = 1;
>> sys1 = tf(num1,den1);
>> num2 = 10;
>> den2 = [1   3   2   0];
>> sys2 = tf(num2,den2);
>> sysg = series(sys1,sys2);
>> num3 = [0.5   1];
>> den3 = 1;
>> sysh = tf(num3,den3);
>> sys = feedback(sysg,sysh)

Transfer function:
       10 s + 50
-----------------------
s^3 + 8 s^2 + 37 s + 50
```

B-10-5. Define the input impedance and feedback impedance as Z_1 and Z_2, respectively, as shown in the figure below.

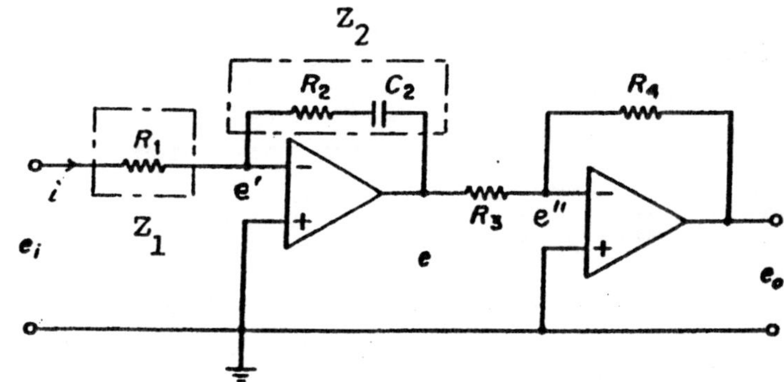

Then

$$Z_1 = R_1 , \qquad Z_2 = R_2 + \frac{1}{C_2 s} = \frac{R_2 C_2 s + 1}{C_2 s}$$

Because each operational amplifier of the system involves negative feedback, the differential input voltage of each amplifier is zero, or e' = 0 and e" = 0. The voltages $E_i(s)$ and $E(s)$ are given by

$$E_i(s) = Z_1 I(s) = R_1 I(s)$$

$$E(s) = - Z_2 I(s) = - \frac{R_2 C_2 s + 1}{C_2 s} I(s)$$

Hence

$$\frac{E(s)}{E_i(s)} = - \frac{R_2 C_2 s + 1}{R_1 C_2 s}$$

Also,

$$\frac{E_o(s)}{E(s)} = - \frac{R_4}{R_3}$$

Therefore,

$$\frac{E_o(s)}{E_i(s)} = \frac{R_4}{R_3} \frac{R_2 C_2 s + 1}{R_1 C_2 s} = \frac{R_2 R_4}{R_1 R_3} \left(1 + \frac{1}{R_2 C_2 s}\right)$$

The control action is proportional plus integral.

B-10-6. If the engine speed increases, the sleeve of the fly-ball governor moves upward. This movement acts as the input to the hydraulic controller. A positive error signal (upward motion of the sleeve) causes the power piston to move downward, reduces the fuel valve opening, and decreases the engine speed. Referring to Figure (a) shown below, a block diagram for the system can be drawn as shown in Figure (b) on next page.

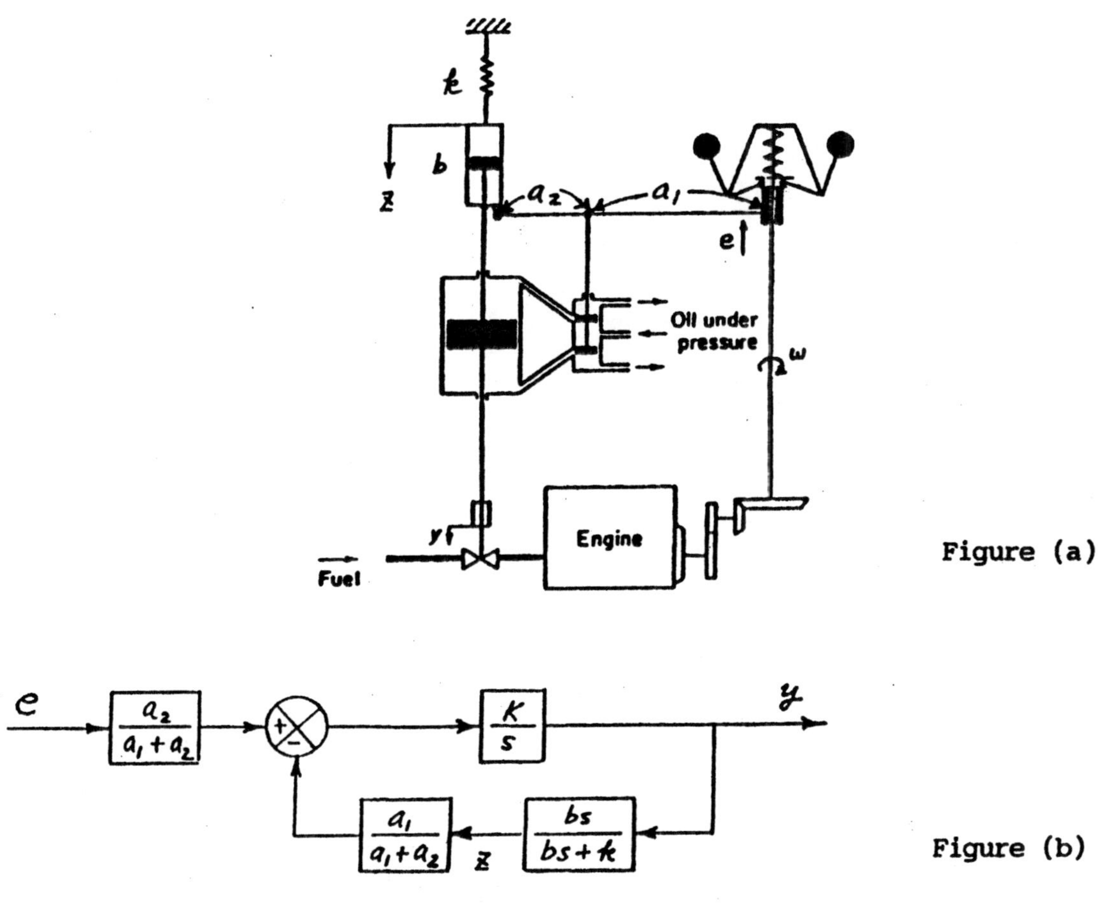

Figure (a)

Figure (b)

From Figure (b) the transfer function Y(s)/E(s) is obtained as

$$\frac{Y(s)}{E(s)} = \frac{a_2}{a_1 + a_2} \frac{\frac{K}{s}}{1 + \frac{K}{s} \frac{a_1}{a_1 + a_2} \frac{bs}{bs + k}}$$

Since such a speed controller is usually designed such that

$$\left| \frac{K}{s} \frac{a_1}{a_1 + a_2} \frac{bs}{bs + k} \right| \gg 1$$

the transfer function Y(s)/E(s) becomes

$$\frac{Y(s)}{E(s)} \doteqdot \frac{a_2}{a_1 + a_2} \frac{a_1 + a_2}{a_1} \frac{bs + k}{bs} = \frac{a_2}{a_1} \left(1 + \frac{k}{bs}\right)$$

Thus, the control action of this speed controller is proportional-plus-integral.

B-10-7. For the first-order system

$$\frac{\theta(s)}{\theta_b(s)} = \frac{1}{Ts + 1}$$

the step response curve is an exponential curve. So the time constant T can be determined from such an exponential curve easily. From Figure 10-98 the time constant T is 2 s.

If this thermometer is placed in a bath, the temperature of which is increasing at a rate of 10°C/min = 1/6°C/s, or

$$\theta_b = \frac{1}{6} t + a$$

where a is a constant, then the steady-state error can be determined as follows: Noting that

$$E(s) = \theta_b(s) - \theta(s) = \theta_b(s)\left[1 - \frac{\theta(s)}{\theta_b(s)}\right]$$

$$= \theta_b(s)(1 - \frac{1}{2s + 1}) = \theta_b(s)\frac{2s}{2s + 1}$$

we obtain

$$e_{ss} = \lim_{s \to 0} sE(s) = \lim_{s \to 0} \frac{2s^2}{2s + 1}\theta_b(s)$$

where

$$\theta_b(s) = \frac{1}{6}\frac{1}{s^2} + \frac{a}{s} = \frac{1 + 6as}{6s^2}$$

Therefore,

$$e_{ss} = \lim_{s \to 0} \frac{2s^2}{2s + 1}\frac{1 + 6as}{6s^2} = \frac{2}{1}\cdot\frac{1}{6} = \frac{1}{3} \text{ °C}$$

Thus, the steady-state error is 1/3°C.

For a second-order system:

$$\frac{\theta(s)}{\theta_b(s)} = \frac{1}{(T_1 s + 1)(T_2 s + 1)}$$

A typical response curve, when this thermometer is placed in a bath held at a constant temperature, is shown in the next page.

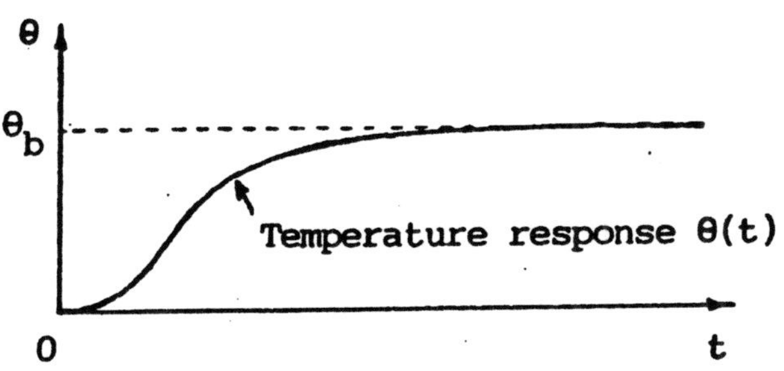

B-10-8. The closed-loop transfer function of the system is

$$\frac{C(s)}{R(s)} = \frac{10(s + 1)}{s^2 + 10s + 10}$$

$$= \frac{10(s + 1)}{(s + 1.1270)(s + 8.8730)}$$

For the unit-step input, we have

$$C(s) = \frac{10(s + 1)}{(s + 1.1270)(s + 8.8730)} \frac{1}{s}$$

$$= \frac{1}{s} + \frac{0.1455}{s + 1.1270} - \frac{1.1455}{s + 8.8730}$$

Hence

$$c(t) = 1 + 0.1455\, e^{-1.1270t} - 1.1455\, e^{-8.8730t}$$

B-10-9. Since M_p is specified as 0.05, we have

$$M_p = e^{-\frac{\zeta\pi}{\sqrt{1-\zeta^2}}} = 0.05$$

or

$$\frac{\zeta\pi}{\sqrt{1-\zeta^2}} = 2.995$$

Rewriting,

$$(\zeta\pi)^2 = (2.995)^2(1 - \zeta^2)$$

Solving for the damping ratio ζ we obtain

$$\zeta = 0.69$$

The settling time t_s is specified as 2 seconds. So we have

$$t_s = \frac{4}{\zeta\omega_n} = 2$$

or

$$\zeta \omega_n = 2$$

Therefore,

$$\omega_n = \frac{2}{\zeta} = \frac{2}{0.69} = 2.90 \text{ rad/s}$$

B-10-10. The closed-loop transfer function of the system is

$$\frac{C(s)}{R(s)} = \frac{100}{s^3 + 2s^2 + 10s + 100}$$

$$= \frac{100}{(s + 4.5815)(s - 1.2907 + j4.4901)(s - 1.2907 - j4.4901)}$$

This system is unstable because two complex-conjugate closed-loop poles are in the right half plane. To visualize the unstable response, we may enter the following MATLAB program into the computer. The resulting unstable response curve is shown below.

To make the system stable, it is necessary to reduce the gain of the system or add an appropriate compensator.

```
>> num = [0   0   0   100];
>> den = [1   2   10   100];
>> t = 0:0.01:6;
>> y = step(num,den,t);
>> plot(t,y)
>> grid
>> title('Unit-Step Response of Unstable System')
>> xlabel('t (sec)')
>> ylabel('c(t)')
```

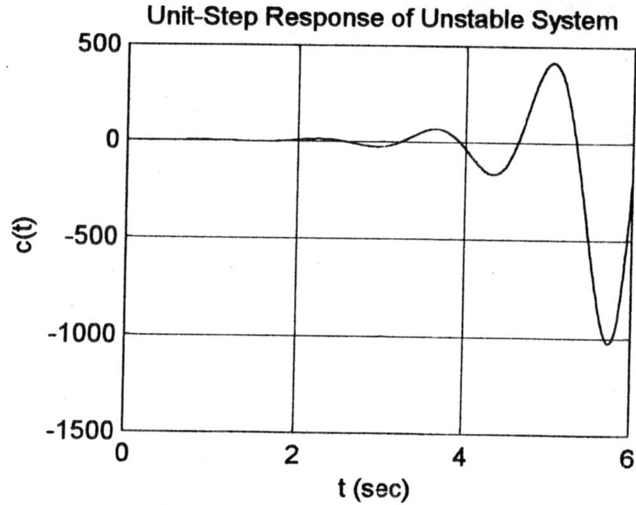

B-10-11.
$$\frac{C(s)}{R(s)} = \frac{16}{s^2 + (0.8 + 16k)s + 16}$$

From the characteristic polynomial, we find

$$\omega_n = 4, \quad 2\zeta\omega_n = 2 \times 0.5 \times 4 = 0.8 + 16k$$

Hence

$$k = 0.2$$

The rise timw t_r is obtained from

$$t_r = \frac{\pi - \beta}{\omega_d}$$

Since

$$\omega_d = \omega_n \sqrt{1 - \zeta^2} = 4\sqrt{1 - 0.25} = 3.46$$

$$\beta = \sin^{-1}\frac{\omega_d}{\omega_n} = \sin^{-1} 0.866 = \frac{\pi}{3}$$

we have

$$t_r = \frac{\pi - \frac{1}{3}\pi}{3.46} = 0.605 \text{ s}$$

The peak time t_p is obtained as

$$t_p = \frac{\pi}{\omega_d} = \frac{3.14}{3.46} = 0.907 \text{ s}$$

The maximum overshoot M_p is

$$M_p = e^{-\frac{\zeta\pi}{\sqrt{1-\zeta^2}}} = e^{-\frac{0.5 \times 3.14}{\sqrt{1-0.25}}} = e^{-1.814} = 0.163$$

The settling time t_s is

$$t_s = \frac{4}{\zeta\omega_n} = \frac{4}{0.5 \times 4} = 2 \text{ s}$$

B-10-12. The closed-loop transfer function for the system is

$$\frac{C(s)}{R(s)} = \frac{K}{2s^2 + s + KK_h s + K} = \frac{0.5K}{s^2 + 0.5(1 + KK_h)s + 0.5K}$$

From this equation, we obtsin

$$\omega_n = \sqrt{0.5K}, \quad 2\zeta\omega_n = 0.5(1 + KK_h)$$

Since the damping ratio ζ is specified as 0.5, we get

$$\omega_n = 0.5(1 + KK_h)$$

Therefore, we have

$$0.5(1 + KK_h) = \sqrt{0.5K}$$

The settling time is specified as

$$t_s = \frac{4}{\zeta \omega_n} = \frac{4}{0.25(1 + KK_h)} = \frac{16}{1 + KK_h} \leqslant 2$$

Since the feedforward transfer function G(s) is

$$G(s) = \frac{\frac{K}{2s + 1}}{1 + \frac{KK_h}{2s + 1}} \frac{1}{s} = \frac{K}{2s + 1 + KK_h} \frac{1}{s}$$

the static velocity error constant K_v is

$$K_v = \lim_{s \to 0} sG(s) = \lim_{s \to 0} s \frac{K}{2s + 1 + KK_h} \frac{1}{s} = \frac{K}{1 + KK_h}$$

This value must be equal to or greater than 50. Hence,

$$\frac{K}{1 + KK_h} \geqslant 50$$

Thus, the conditions to be satisfied can be summarized as follows:

$$0.5(1 + KK_h) = \sqrt{0.5K} \quad (1)$$

$$\frac{16}{1 + KK_h} \leqslant 2 \quad (2)$$

$$\frac{K}{1 + KK_h} \geqslant 50 \quad (3)$$

$$0 < K_h < 1 \quad (4)$$

From Equations (1) and (2), we gwt

$$8 \leqslant 1 + KK_h = \sqrt{2K}$$

or

$$32 \leqslant K$$

From Equation (3) we obtain

$$\frac{K}{50} \geqslant 1 + KK_h = \sqrt{2K}$$

or

$$K \geqslant 5000$$

If we choose K = 5000, then we get

$$1 + KK_h = \sqrt{2K} = 100$$

or

$$K_h = \frac{99}{5000} = 0.0198$$

Thus, we determined a set of values of K and K_h as follows:

$$K = 5000, \quad K_h = 0.0198$$

With these values of K and K_h, all specifications are satisfied.

B-10-13.

$$\frac{C(s)}{R(s)} = \frac{K_p(1 + T_d s)}{Js^2 + K_p(1 + T_d s)}$$

Since $R(s) = 1/s^2$, the output C(s) is obtained as

$$C(s) = \frac{K_p + K_p T_d s}{Js^2 + K_p T_d s + K_p} \frac{1}{s^2} = \frac{1}{s^2} - \frac{1}{s^2 + \frac{K_p T_d}{J} s + \frac{K_p}{J}}$$

Since the system is underdamped, C(s) can be written as

$$C(s) = \frac{1}{s^2} - \frac{1}{\left(s + \frac{K_p T_d}{2J}\right)^2 + \frac{K_p}{J} - \frac{K_p^2 T_d^2}{4J^2}}$$

$$= \frac{1}{s^2} - \frac{1}{\sqrt{\frac{K_p}{J} - \frac{K_p^2 T_d^2}{4J^2}}} \frac{\sqrt{\frac{K_p}{J} - \frac{K_p^2 T_d^2}{4J^2}}}{\left(s + \frac{K_p T_d}{2J}\right)^2 + \left(\sqrt{\frac{K_p}{J} - \frac{K_p^2 T_d^2}{4J^2}}\right)^2}$$

The inverse Laplace transform of C(s) gives

$$c(t) = t - \frac{1}{\sqrt{\frac{K_p}{J} - \frac{K_p^2 T_d^2}{4J^2}}} e^{-\frac{K_p T_d}{2J} t} \sin\sqrt{\frac{K_p}{J} - \frac{K_p^2 T_d^2}{4J^2}}\, t$$

The steady-state error e_{ss} for a unit ramp input is

$$e_{ss} = \lim_{t \to \infty} [r(t) - c(t)] = \lim_{t \to \infty} [t - c(t)]$$

$$= \lim_{t \to \infty} \left(\frac{1}{\sqrt{\frac{K_p}{J} - \frac{K_p^2 T_d^2}{4J^2}}} e^{-\frac{K_p T_d}{2J} t} \sin \sqrt{\frac{K_p}{J} - \frac{K_p^2 T_d^2}{4J^2}} \, t \right) = 0$$

The steady-state error can also be obtained by use of the final value theorem. Since the error signal E(s) is

$$E(s) = R(s) - C(s) = R(s) \left[1 - \frac{C(s)}{R(s)} \right] = \frac{1}{s^2} \left[1 - \frac{K_p(1 + T_d s)}{Js^2 + K_p(1 + T_d s)} \right]$$

we obtain the steady-state error e_{ss} as

$$e_{ss} = \lim_{t \to \infty} e(t) = \lim_{s \to 0} sE(s) = \lim_{s \to 0} \frac{s\, Js^2}{s^2(Js^2 + K_p + T_d K_p s)} = 0$$

B-10-14. The characteristic equation is

$$\frac{K}{s(s + 1)(s + 5)} + 1 = 0$$

or

$$s^3 + 6s^2 + 5s + K = 0$$

The Routh array for this equation is

$$\begin{array}{cc} 1 & 5 \\ 6 & K \\ \frac{30 - K}{6} & 0 \\ K & \end{array}$$

For the system to be stable, there should be no sign changes in the first column. This requires

$$30 - K > 0, \qquad K > 0$$

Hence, we get the range of gain K for stability to be

$$30 > K > 0$$

B-10-15. Since the system is of higher order (5th order), it is easier to find the range of gain K for stability by first plotting the root loci and then finding critical points (for stability) on the root loci. The open-loop transfer function G(s) can be written as

$$G(s) = \frac{K(s^2 + 2s + 4)}{s(s + 4)(s + 6)(s^2 + 1.4s + 1)}$$

$$= \frac{K(s^2 + 2s + 4)}{s^5 + 11.4s^4 + 39s^3 + 43.6s^2 + 24s}$$

The MATLAB program given next will generate a plot of the root loci for the system. The resulting root-locus plot is shown below the MATLAB program.

```
>> num = [0    0    0    1    2    4];
>> den = [1    11.4    39    43.6    24    0];
>> rlocus(num,den)
>> v = [-8    2    -5    5]; axis(v); axis('equal')
>> title('Root-Locus Plot')
```

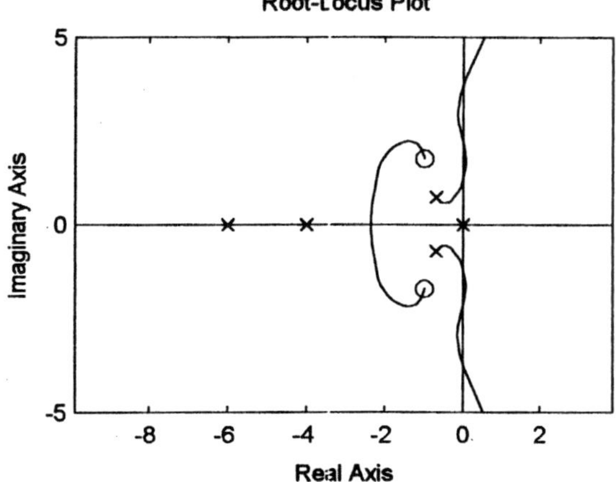

Based on this plot, it can be seen that the system is conditionally stable. All critical points for stability lie on the $j\omega$ axis.

To determine the crossing points of the root loci with the $j\omega$ axis, substitute $s = j\omega$ into the characteristic equation which is

$$s^5 + 11.4s^4 + 39s^3 + 43.6s^2 + 24s + K(s^2 + 2s + 4) = 0$$

Then

$$(j\omega)^5 + 11.4(j\omega)^4 + 39(j\omega)^3 + (43.6 + K)(j\omega)^2 + (24 + 2K)j\omega$$
$$+ 4K = 0$$

This equation can be rewritten as

$$[11.4\,\omega^4 - (43.6 + K)\omega^2 + 4K] + j[\omega^5 - 39\omega^3 + (24 + 2K)\omega] = 0$$

By equating the real part and imaginary part equal to zero, respectively, we obtain

$$11.4\,\omega^4 - (43.6 + K)\,\omega^2 + 4K = 0 \qquad (1)$$

$$\omega^5 - 39\,\omega^3 + (24 + 2K)\omega = 0 \qquad (2)$$

Equation (2) can be written as

$$\omega = 0$$

or

$$\omega^4 - 39\,\omega^2 + 24 + 2K = 0 \qquad (3)$$

From Equation (3) we obtain

$$K = \frac{-\omega^4 + 39\,\omega^2 - 24}{2} \qquad (4)$$

By substituting Equation (4) into Equation (1), we get

$$11.4\,\omega^4 - [43.6 + \tfrac{1}{2}(-\omega^4 + 39\,\omega^2 - 24)]\,\omega^2 - 2\,\omega^4$$

$$+ 78\,\omega^2 - 48 = 0$$

which can be simplified to

$$\omega^6 - 20.2\,\omega^4 + 92.8\,\omega^2 - 96 = 0$$

The roots of this last equation can be easily obtained by use of the MATLAB program as given below.

```
>> a = [1    0    -20.2    0    92.8    0    -96];
>> roots(a)

ans =

   3.7553
  -3.7553
   2.1509
   1.2130
  -2.1509
  -1.2130
```

The root-locus branch in the upper half plane that goes to infinity crosses the $j\omega$ axis at $\omega = 1.2130$, $\omega = 2.1509$, and $\omega = 3.7553$. The gain values at these crossing points are obtained as follows:

$$K = \frac{-1.2130^4 + 39 \times 1.2130^2 - 24}{2} = 15.61 \qquad \text{for } \omega = 1.2130$$

$$K = \frac{-2.1509^4 + 39 \times 2.1509^2 - 24}{2} = 67.51 \qquad \text{for } \omega = 2.1509$$

$$K = \frac{-3.7553^4 + 39 \times 3.7553^2 - 24}{2} = 163.56 \qquad \text{for } \omega = 3.7553$$

Based on the K values above, we obtain the range of gain K for stability as follows: The system is stable if

$$15.61 > K > 0$$

$$163.56 > K > 67.51$$

B-10-16. A MATLAB program to plot the root loci and asymptotes for the following system

$$G(s)H(s) = \frac{K}{s(s + 0.5)(s^2 + 0.6s + 10)}$$

$$= \frac{K}{s^4 + 1.1s^3 + 10.3s^2 + 5s}$$

is given below and the resulting root-locus plots are shown on the next page.

Note that the equation for the asymptotes is

$$G_a(s)H_a(s) = \frac{K}{(s + 0.275)^4}$$

$$= \frac{K}{s^4 + 1.1s^3 + 0.4538s^2 + 0.08319s + 0.005719}$$

```
>> num = [0    0    0    0    1];
>> den = [1    1.1    10.3    5    0];
>> numa = [0    0    0    0    1];
>> dena = [1    1.1    0.4538    0.08319    0.005719];
>> subplot(221); rlocus(num,den);
>> v = [-5    5    -5    5]; axis(v); axis('equal')
>> subplot(222); rlocus(num,den)
>> hold
Current plot held
>> rlocus(numa,dena);
>> v = [-5    5    -5    5]; axis(v); axis('equal')
>> title('Plot of Root Loci and Asymptotes')
```

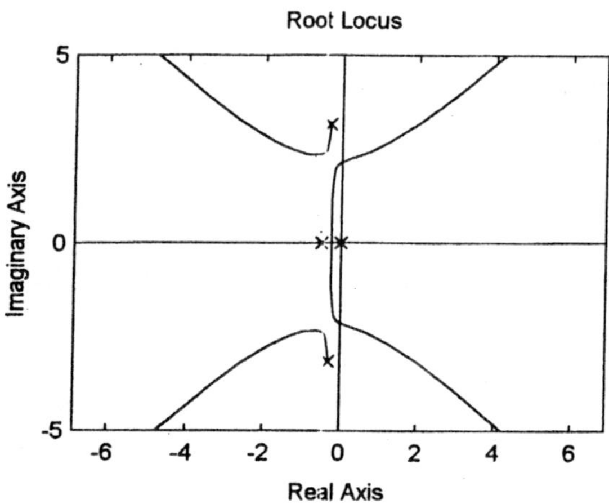

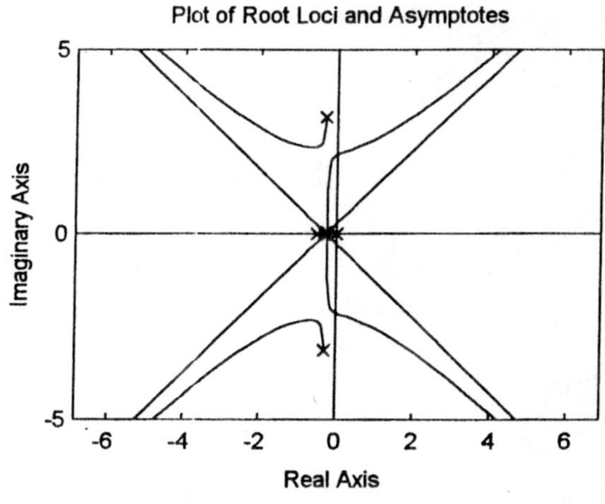

B-10-17. The open-loop transfer function G(s) is

$$G(s) = K \frac{s+1}{s+5} \frac{2}{s^2(s+2)}$$

$$= \frac{K2(s+1)}{s^4 + 7s^3 + 10s^2}$$

The following MATLAB program will generate a root-locus plot. The resulting plot is shown below.

```
>> num = [0   0   0   1   1];
>> den = [1   7   10   0   0];
>> rlocus(num,den)
>> v = [-4   2   -3   3]; axis(v); axis('equal')
>> title('Root-Locus Plot')
```

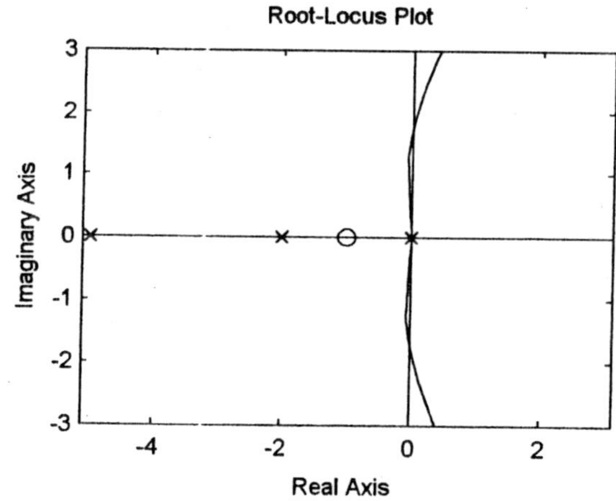

From the plot we find that the critical value of gain K for stability corresponds to the crossing point of the root locus branch that goes to infinity and the imaginary axis. Hence, we first find the crossing frequency and then find the corresponding gain value.

The characteristic equation for this system is

$$s^4 + 7s^3 + 10s^2 + 2Ks + 2K = 0$$

By substituting $s = j\omega$ into the characteristic equation, we obtain

$$(j\omega)^4 + 7(j\omega)^3 + 10(j\omega)^2 + 2K(j\omega) + 2K = 0$$

which can be rewritten as

$$(\omega^4 - 10\omega^2 + 2K) + j\omega(-7\omega^2 + 2K) = 0$$

By equating the real part and imaginary part of this last equation to zero, respectively, we get

$$\omega^4 - 10\omega^2 + 2K = 0 \qquad (1)$$

$$\omega(-7\omega^2 + 2K) = 0 \qquad (2)$$

Equation (2) can be rewritten as

$$\omega = 0$$

or

$$-7\omega^2 + 2K = 0 \qquad (3)$$

By substituting Equation (3) into Equation (1), we find

$$\omega^4 - 10\omega^2 + 7\omega^2 = 0$$

or

$$\omega^4 - 3\omega^2 = 0$$

which yields

$$\omega = 0, \quad \omega = 0, \quad \omega = \sqrt{3}, \quad \omega = -\sqrt{3}$$

Since $\omega = \sqrt{3}$ is the crossing frequency with the $j\omega$ axis, by substituting $\omega = \sqrt{3}$ into Equation (3) we obtain the critical value of gain K for stability.

$$K = 3.5\omega^2 = 3.5 \times 3 = 10.5$$

Hence the stability range for K is

$$10.5 > K > 0$$

B-10-18. The angle deficiency is

$$180° - 120° - 120° = -60°$$

A lead compensator can contribute 60°. Let us choose the zero of the lead compensator at s = -1. Then, to obtain phase lead angle of 60°, the pole of the compensator must be located at s = -4. Thus,

$$G_c(s) = K \frac{s + 1}{s + 4}$$

The gain K can be determined from the magnitude condition.

$$\left| K \frac{s + 1}{s + 4} \frac{1}{s^2} \right|_{s = -1 + j\sqrt{3}} = 1$$

or

$$K = \left| \frac{(s + 4)s^2}{s + 1} \right|_{s = -1 + j\sqrt{3}} = 8$$

Hence the lead compensator becomes as follows:

$$G_c(s) = 8 \frac{s + 1}{s + 4}$$

The feedforward transfer function is

$$G_c(s)G(s) = \frac{8s + 8}{s^3 + 4s^2}$$

The following MATLAB program will generate a root-locus plot. The resulting plot is shown below the MATLAB program.

```
>> num = [0    0    8    8];
>> den = [1    4    0    0];
>> rlocus(num,den)
>> v = [-6    2    -4    4]; axis(v); axis('equal')
```

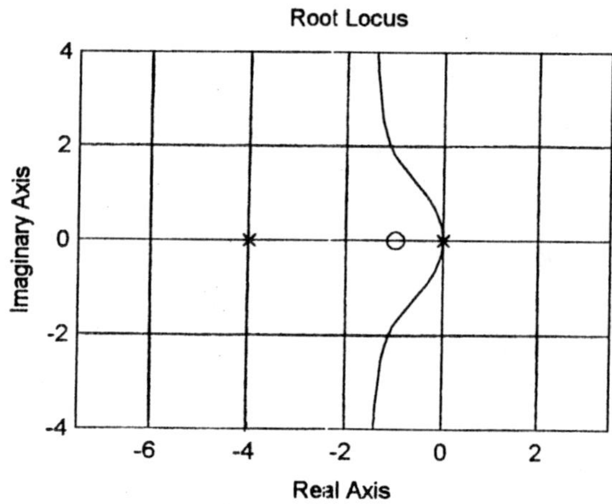

Note that the closed-loop transfer function is

$$\frac{C(s)}{R(s)} = \frac{8s + 8}{s^3 + 4s^2 + 8s + 8}$$

The closed-loop poles are located at $s = -1 \pm j\sqrt{3}$ and $s = -2$.

B-10-19. The MATLAB program given below generates a root-locus plot for the given system. The resulting plot is shown below the program.

```
>> num = [0    0    0    1];
>> den = [1    5    4    0];
>> rlocus(num,den)
>> v = [-6    4    -5    5]; axis(v); axis('equal')
```

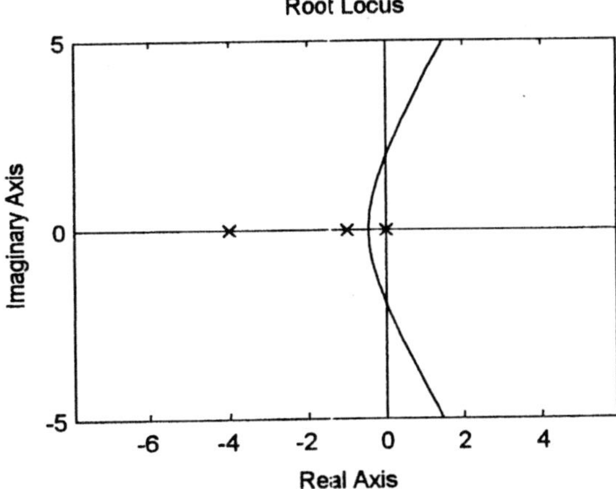

Note that constant-ζ points ($0 < \zeta < 1$) lie on a straight line having angle θ from the $j\omega$ axis as shown in the figure below.
From the figure we obtain

$$\sin\theta = \frac{\zeta\omega_n}{\omega_n} = \zeta$$

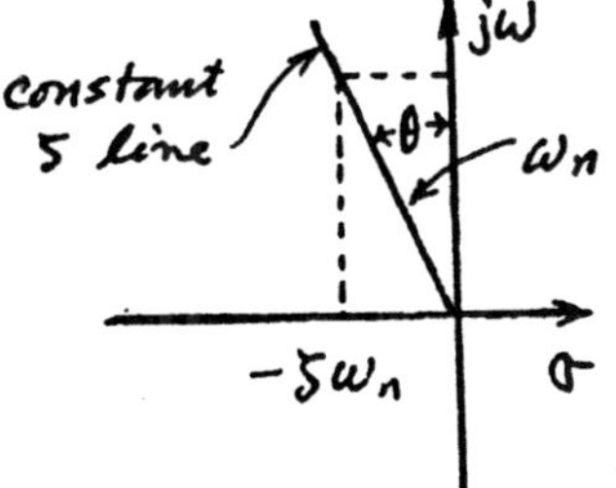

Note also that $\zeta = 0.6$ line can be defined by

$$s = -0.75a + ja$$

where a is a variable ($0 < a < \infty$). To find the value of K such that the damping ratio ζ of the dominant closed-loop poles is 0.6 can be found by finding the intersection of the line $s = -0.75a + ja$ and the root locus. The intersection point can be determined by solving the following simultaneous equations for a.

$$s = -0.75a + ja \tag{1}$$

$$s(s + 1)(s + 4) + K = 0 \tag{2}$$

By substituting Equation (1) into Equation (2),

$$(-0.75a + ja)(-0.75a + ja + 1)(-0.75a + ja + 4) + K = 0$$

which can be rewritten as

$$(1.8281a^3 - 2.1875a^2 - 3a + K) + j(0.6875a^3 - 7.5a^2 + 4a) = 0$$

By equating the real part and imaginary part of this last equation to zero, respectively, we obtain

$$1.8281a^3 -2.1875a^2 - 3a + K = 0 \tag{3}$$

$$0.6875a^3 - 7.5a^2 + 4a = 0 \tag{4}$$

Equation (4) can be rewritten as

$$a = 0$$

or

$$0.6875a^2 - 7.5a + 4 = 0$$

which can be written as

$$a^2 - 10.9091a + 5.8182 = 0$$

or

$$(a - 0.5623)(a - 10.3468) = 0$$

Hence

$$a = 0.5623 \qquad \text{or} \qquad a = 10.3468$$

From Equation (3) we find

$$K = -1.8281a^3 + 2.1875a^2 + 3a = 2.0535 \qquad \text{for } a = 0.5623$$

$$K = -1.8281a^3 + 2.1875a^2 + 3a = -1759.74 \qquad \text{for } a = 10.3468$$

Since the K value is positive for $a = 0.5623$ and negative for $a = 10.3468$, we choose $a = 0.5623$. The required gain K is 2.0535.

Since the characteristic equation with $K = 2.0535$ is

$$s(s + 1)(s + 4) + 2.0535 = 0$$

or

$$s^3 + 5s^2 + 4s + 2.0535 = 0$$

the closed-loop poles can be obtained by use of the following MATLAB program.

```
>> p = [1   5   4   2.0535];
>> roots(p)

ans =

   -4.1565
   -0.4217 + 0.5623i
   -0.4217 - 0.5623i
```

Thus, the closed-loop poles are located at

$$s = -0.4217 \pm j0.5623, \quad s = -4.1565$$

The unit-step response of the system with $K = 2.0535$ can be obtained by entering the following MATLAB program into the computer. The resulting unit-step response curve is shown below the MATLAB program.

```
>> num = [0   0   0   2.0535];
>> den = [1   5   4   2.0535];
>> t = 0:0.01:20;
>> y = step(num,den,t);
>> plot(t,y)
>> grid
>> title('Unit-Step Response')
>> xlabel('t (sec)')
>> ylabel('Output')
```

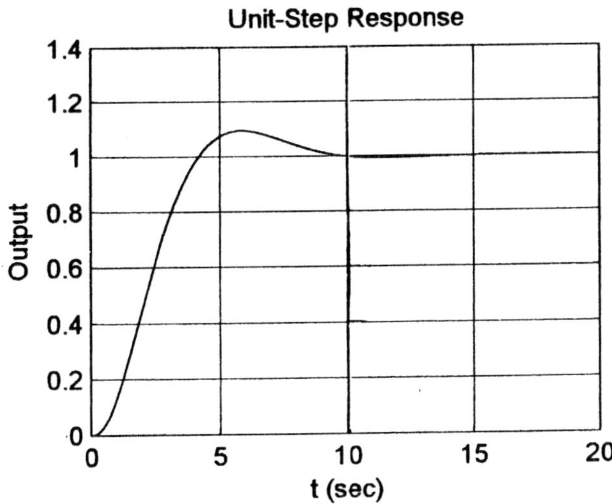

B-10-20. The closed-loop transfer function of the system is

$$\frac{C(s)}{R(s)} = \frac{K(s + a)(s + b)}{s(s^2 + 1) + K(s + a)(s + b)}$$

$$= \frac{K(s^2 + as + bs + ab)}{s^3 + s + K(s^2 + as + bs + ab)}$$

Since the dominant closed-loop poles need be located at $s = -1 \pm j1$, the characteristic equation must be divisible by

$$(s + 1 + j1)(s + 1 - j1) = s^2 + 2s + 2$$

Hence

$$s^3 + Ks^2 + (1 + aK + bK)s + abK = (s^2 + 2s + 2)(s + \alpha)$$

where $s = -\alpha$ is the unknown third pole. By dividing the left side of this last equation by $s^2 + 2s + 2$, we obtain

$$s^3 + Ks^2 + (1 + aK + bK)s + abK = (s^2 + 2s + 2)(s + K - 2)$$

$$+ (aK + bK - 2K + 3)s + Kab - 2(K - 2)$$

The remainder of division must be zero. Hence we set

$$aK + bK - 2K + 3 = 0$$

$$Kab - 2(K - 2) = 0$$

Since a is specified as 0.5, by substituting $a = 0.5$ into these two equations, we obtain

$$bK = 1.5K - 3 \qquad (1)$$

$$0.5Kb - 2(K - 2) = 0 \qquad (2)$$

By substituting Equation (1) into Equation (2), we have

$$0.5(1.5K - 3) - 2(K - 2) = 0$$

or

$$K = 2$$

Then, by substituting $K = 2$ into Equation (1), we get

$$2b = 1.5 \times 2 - 3 = 0$$

Hence

$$b = 0$$

The PID controller with $K = 2$ and $b = 0$ becomes

$$G_c(s) = K\frac{(s + a)(s + b)}{s} = K\frac{(s + 0.5)s}{s} = K(s + 0.5)$$

Thus, the controller becomes a PD controller. The open-loop transfer function becomes

$$G_c(s)G(s) = \frac{K(s + 0.5)}{s^2 + 1}$$

The closed-loop transfer function (with b = 0 and K = 2) becomes as follows:

$$\frac{C(s)}{R(s)} = \frac{2(s + 0.5)}{s^2 + 2s + 2}$$

$$= \frac{2s + 1}{(s + 1 + j1)(s + 1 - j1)}$$

The root-locus plot for the designed system can be obtained by enterring the following MATLAB program into the computer.

```
>> num = [0   1   0.5];
>> den = [1   0   1];
>> rlocus(num,den)
>> v = [-2   1   -1.5   1.5]; axis(v); axis('equal')
```

The resulting root-locus plot is shown below.

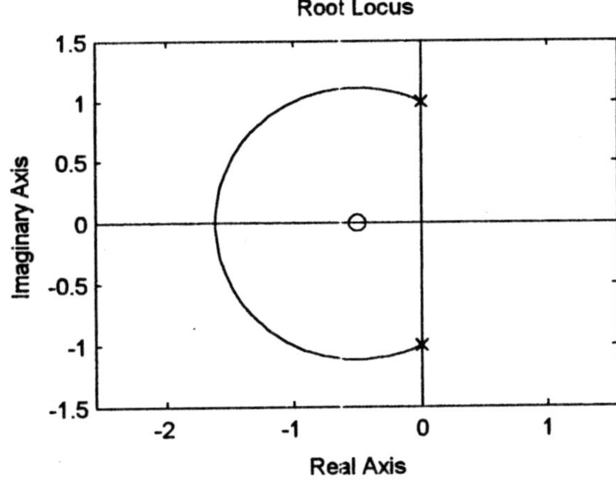

CHAPTER 11

B-11-1.

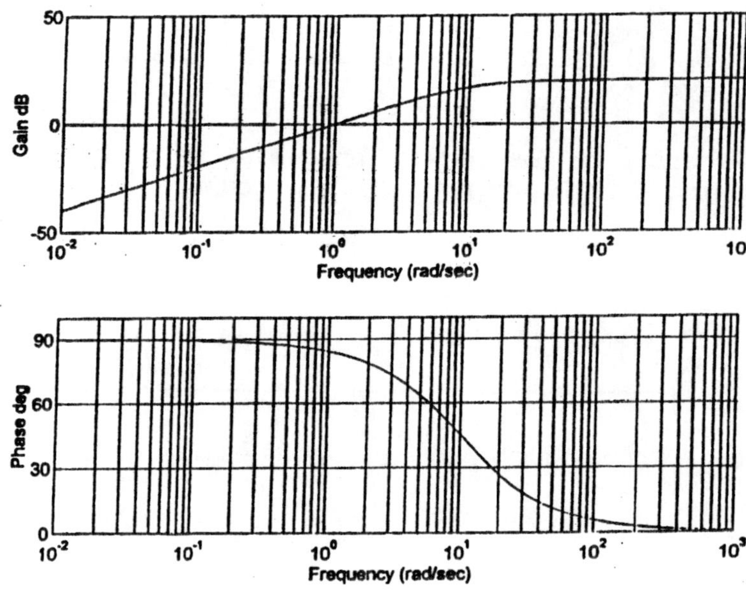

B-11-2.

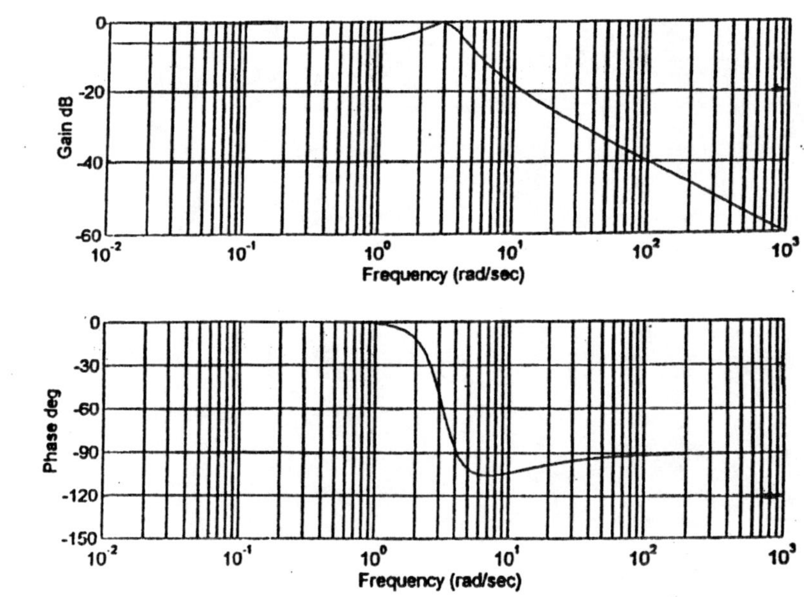

B-11-3. A Bode diagram of the PI controller is shown in Figure (a).

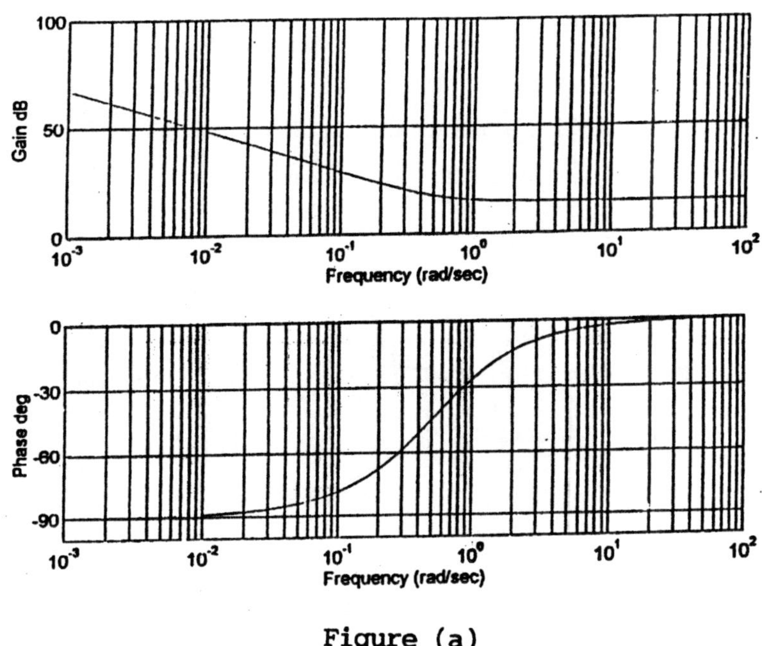

Figure (a)

A Bode diagram of the PD controller is shown in Figure (b).

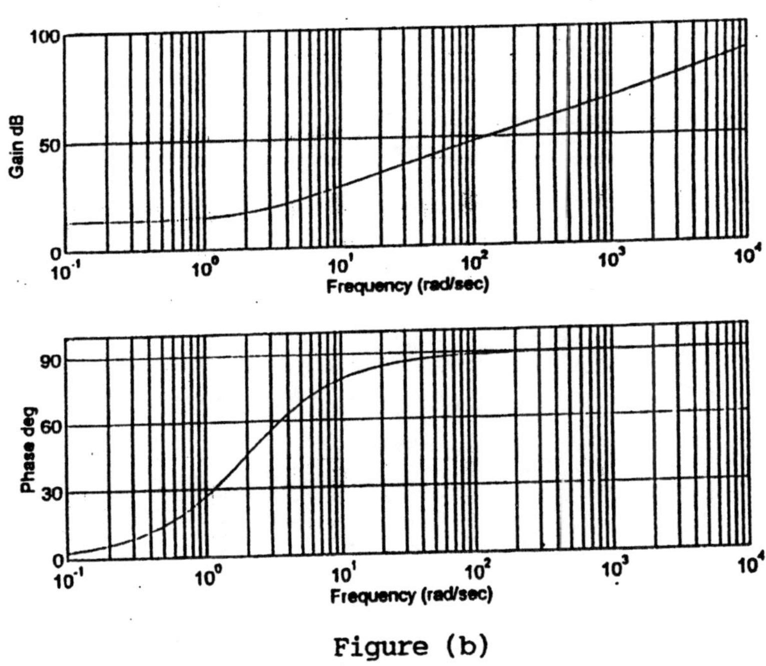

Figure (b)

B-11-4.

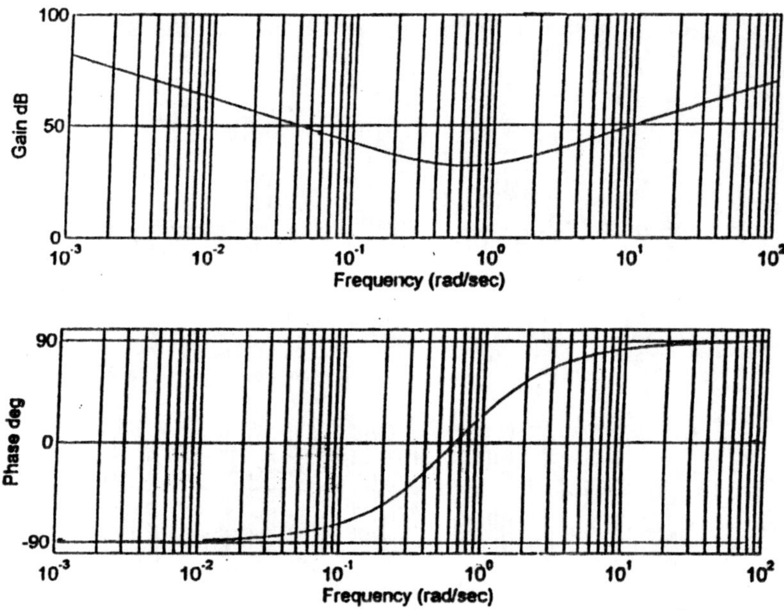

B-11-5.

Lead network

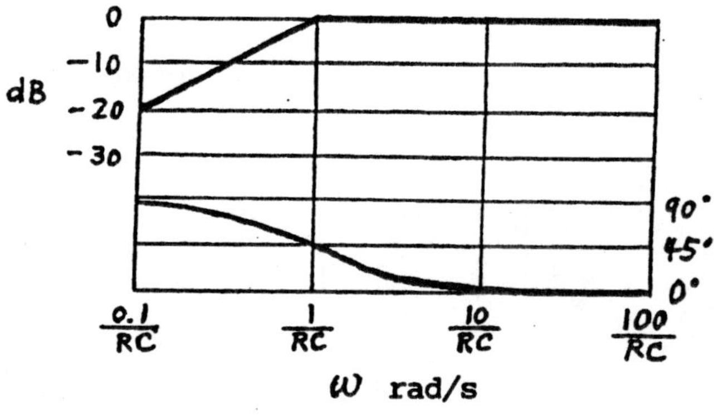

Lag network

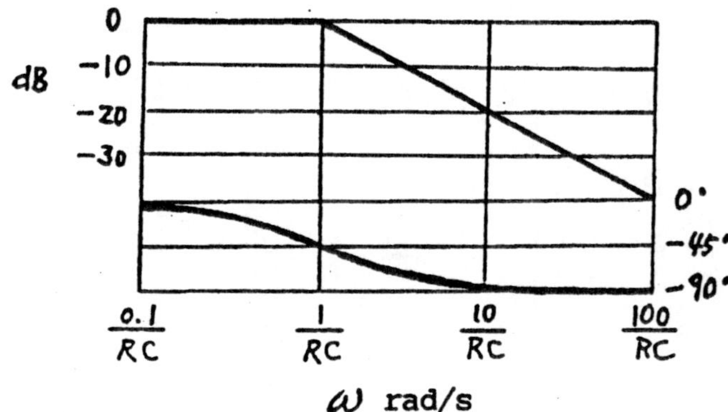

B-11-6. The equation of motion for the system is

$$b(\dot{x} - \ell\dot{\theta}) = k\ell\theta$$

or

$$\ell\dot{\theta} + \frac{k}{b}\ell\theta = \dot{x}$$

The $\mathcal{L}$- transform of this equation, using zero initial conditions, gives

$$\left(\ell s + \frac{k}{b}\ell\right)\Theta(s) = sX(s)$$

Hence

$$\frac{\Theta(s)}{X(s)} = \frac{1}{\ell}\frac{s}{s + (k/b)}$$

Notice that this system is a differentiating system.

Since

$$G(j\omega) = \frac{\Theta(j\omega)}{X(j\omega)} = \frac{1}{\ell}\frac{j\omega}{j\omega + (k/b)}$$

we obtain

$$\left|G(j\omega)\right| = \frac{1}{\ell}\frac{\omega}{\sqrt{\left(\frac{k}{b}\right)^2 + \omega^2}}$$

and

$$\angle G(j\omega) = 90° - \tan^{-1}\frac{\omega b}{k}$$

The steady-state output $\theta_{ss}(t)$ is therefore given by

$$\theta_{ss}(t) = \frac{1}{\ell}\frac{\omega X}{\sqrt{\left(\frac{k}{b}\right)^2 + \omega^2}}\sin(\omega t + 90° - \tan^{-1}\frac{\omega b}{k})$$

Next, substituting ℓ = 0.1 m, k = 2 N/m, and b = 0.2 N-s/m into $G(j\omega)$ gives

$$G(j\omega) = 10\frac{j\omega}{j\omega + 10}$$

A Bode diagram of $G(j\omega)$ is shown on next page.

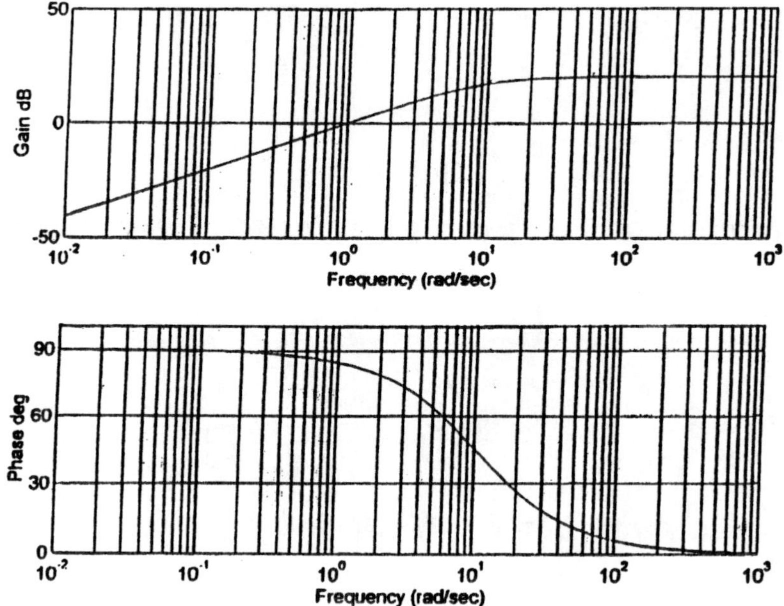

B-11-7. **Noting that**

$$G(j\omega) = \frac{\omega_n^2}{(j\omega)^2 + 2\zeta\omega_n(j\omega) + \omega_n^2}$$

$$= \frac{1}{\left(j\dfrac{\omega}{\omega_n}\right)^2 + 2\zeta\left(j\dfrac{\omega}{\omega_n}\right) + 1}$$

we have

$$\left|G(j\omega_n)\right| = \left|\frac{1}{-1 + 2\zeta j + 1}\right| = \frac{1}{2\zeta}$$

B-11-8. **A possible MATLAB program for obtaining a Bode diagram of the** given G(s) is shown below. The resulting Bode diagram is shown on next page.

```
>> num = [0    0    0    320    640];
>> den = [1    9    72    64    0];
>> w = logspace(-2,3,100);
>> bode(num,den,w)
>> grid
```

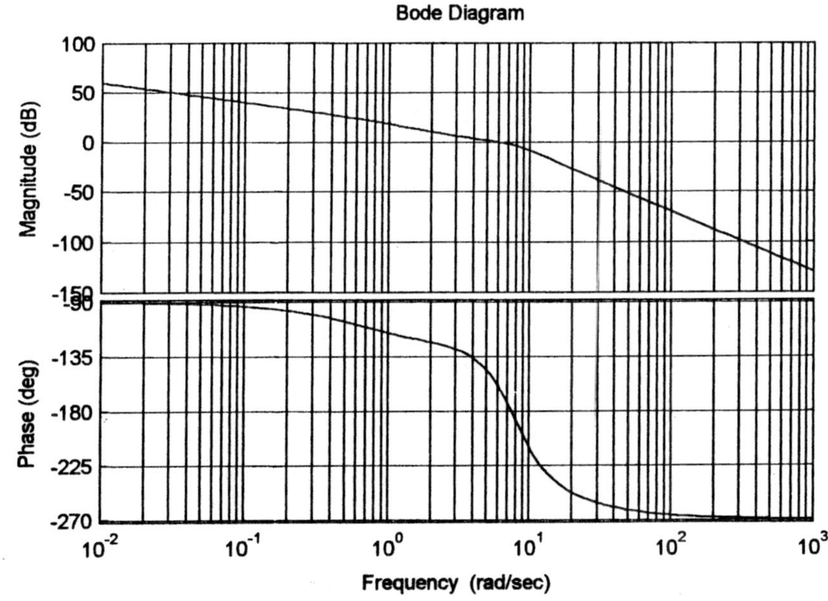

B-11-9. **A possible MATLAB program to obtain a Bode diagram of the given G(s) is shown below.** The resulting Bode diagram is also shown below.

```
>> num = [0    20    20    10];
>> den = [1    11    10     0];
>> w = logspace(-2,3,100);
>> bode(num,den,w)
>> grid
```

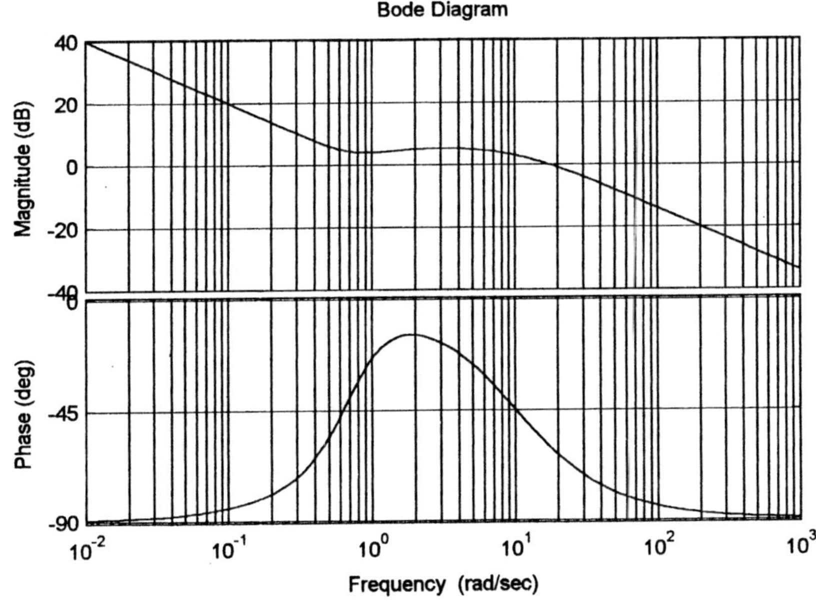

B-11-10. A possible MATLAB program for obtaining a Nyquist plot of the given G(s) is shown below. Note that to plot $G(j\omega)$ locus only for $\omega > 0$, we use the following command:

```
[re,im,w] = nyquist(num,den,w);

plot(re,im)
```

The resulting Nyquist plot is shown below the MATLAB program.

```
>> num = [0    0    0    1];
>> den = [1    0.8    1    0];
>> w = 0.1:0.1:100;
>> [re,im,w] = nyquist(num,den,w);
>> plot(re,im)
>> v = [-3    3    -4    2]; axis(v); axis('equal')
>> grid
>> title('Nyquist Plot')
>> xlabel('Real Axis')
>> ylabel('Imaginary Axis')
```

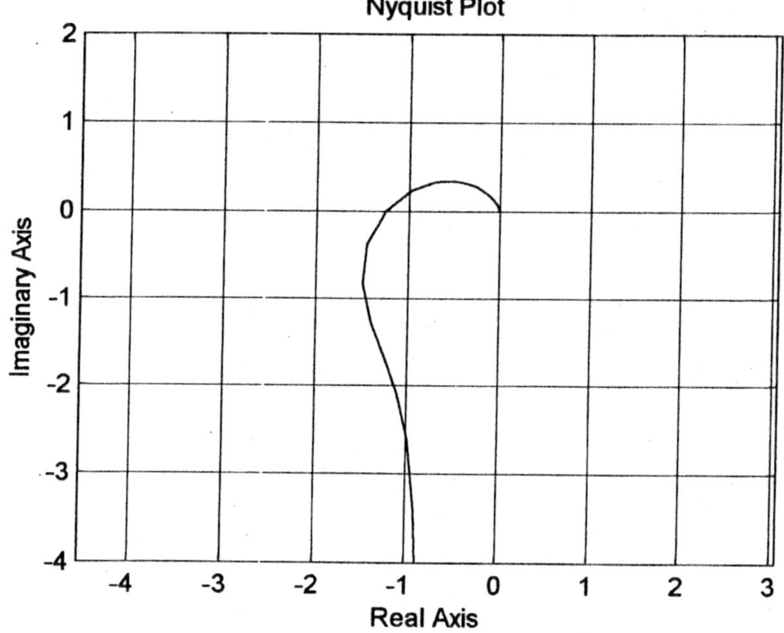

Since none of the open-loop poles lie in the right-half s plane and the $G(j\omega)$ locus encircles the $-1 + j0$ point twice clockwise if $G(j\omega)$ locus is plotted from $\omega = -\infty$ to $\omega = \infty$, the closed-loop system is unstable.

B-11-11. A possible MATLAB program for obtaining a Nyquist plot of the given G(s) is shown below. The resulting Nyquist plot is also shown below. Since none of the open-loop poles lie in the right-half s plane and from the Nyquist plot it can be seen that the $G(j\omega)$ locus does not encircle the $-1 + j0$ point, the system is stable.

```
>> num = [0    0    0    20    20];
>> den = [1    7    20    50    0];
>> w = 0.1:0.1:100;
>> [re,im,w] = nyquist(num,den,w);
>> plot(re,im)
>> v = [-1.5    1.5    -2.5    0.5]; axis(v); axis('equal')
>> grid
>> title('Nyquist Plot')
>> xlabel('Real Axis')
>> ylabel('Imaginary Axis')
```

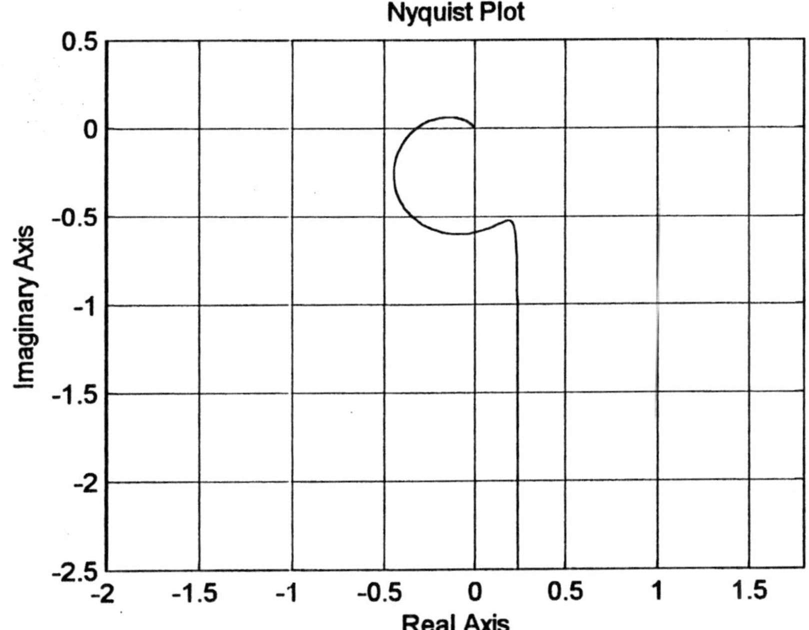

B-11-12. A possible MATLAB program for obtaining a Nyquist plot of the given G(s) is shown on next page. The resulting Nyquist plot is also shown on next page.

Note that there are two open-loop poles in the right-half s plane, because

$$s^3 + 0.2s^2 + s + 1$$

$$= (s + 0.7246)(s - 0.2623 + j1.1451)(s - 0.2623 - j1.1451)$$

```
>> num = [0    1    2    1];
>> den = [1    0.2    1    1];
>> w =0:0.005:10;
>> [re,im,w] = nyquist(num,den,w);
>> plot(re,im)
>> v = [-3    3    -3    3]; axis(v); axis('equal')
>> grid
>> title('Nyquist Plot')
>> xlabel('Real Axis')
>> ylabel('Imaginary Axis')
```

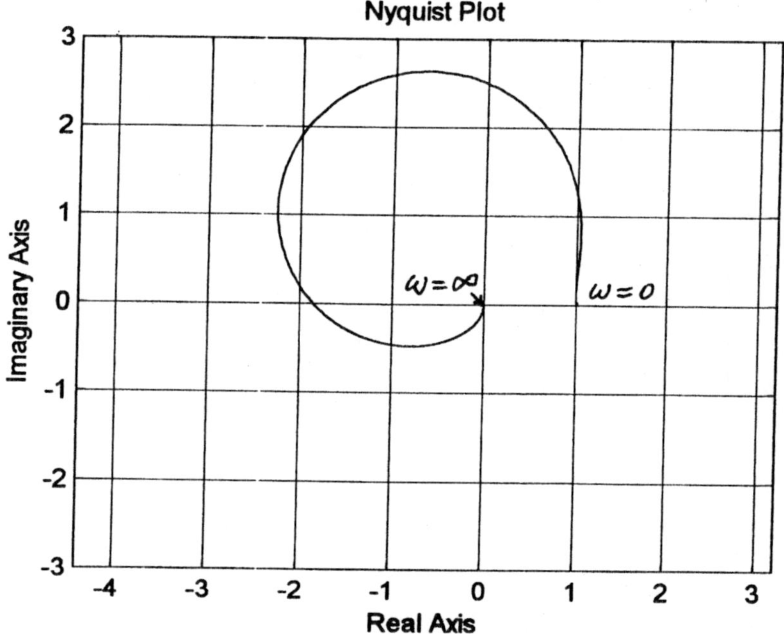

From the plot, it is seen that the $G(j\omega)$ locus encircles the $-1 + j0$ point twice as ω is varied from $\omega = -\infty$ to $\omega = 0$ to $\omega = \infty$. Referring to the Nyquist stability criterion, we have

N = number of clockwise encirclement of the $-1 + j0$ point = -2

P = number of poles of G(s) in the right-half s plane = 2

Then

Z = number of zeros of 1 + G(s) in the right-half s plane

$$= N + P = -2 + 2 = 0$$

Thus, there are no closed-loop poles in the right-half s plane and the closed-loop system is stable.

B-11-13. A closed-loop system with the following open-loop transfer function

$$G(s)H(s) = \frac{K}{s^2(T_1 s + 1)} \qquad (T_1 > 0) \qquad (1)$$

is unstable, while a closed-loop system with the following open-loop transfer function is stable.

$$G(s)H(s) = \frac{K(T_2 s + 1)}{s^2(T_1 s + 1)} \qquad (T_2 > T_1 > 0) \qquad (2)$$

Nyquist plots of the preceding two systems are shown below. Note that $G(j\omega)H(j\omega)$ loci start from negative infinity on the real axis ($\omega = 0$) and approach the origin ($\omega = \infty$). The system with the open-loop transfer function given by Equation (1) encircles the $-1 + j0$ point twice clockwise. The system is unstable. The system with open-loop transfer function given by equation (2) does not encircle the $-1 + j0$ point. Hence, this system is stable.

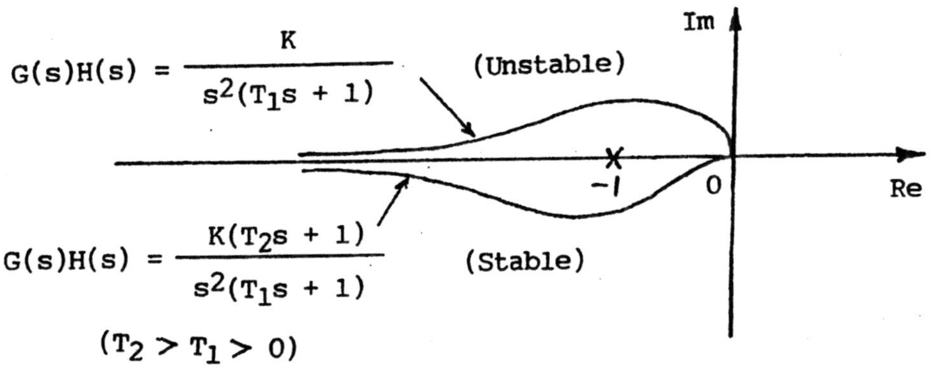

B-11-14. For this system

$$G(j\omega) = K \frac{e^{-j\omega}}{j\omega + 1}$$

By setting K = 1, we draw a Nyquist diagram as shown on next page. Note that

$$\angle e^{-j\omega} = -\omega \text{ (rad)} = -57.3^\circ \omega$$

The Nyquist locus crosses the negative real axis at $\sigma = -0.442$. Hence for stability, we require

$$\frac{1}{0.442} > K > 0$$

or

$$2.262 > K > 0$$

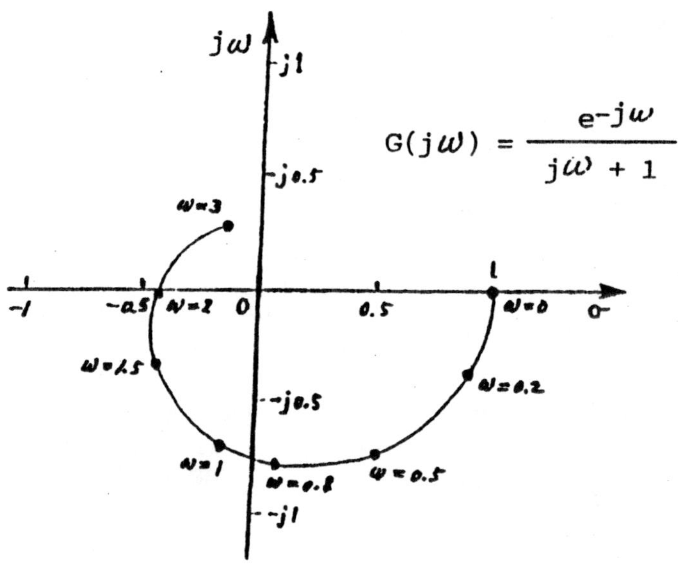

The same result can also be obtained analytically. Since

$$G(j\omega) = \frac{Ke^{-j\omega}}{j\omega + 1} = \frac{K(\cos\omega - j\sin\omega)(1 - j\omega)}{(1 + j\omega)(1 - j\omega)}$$

$$= \frac{K}{1 + \omega^2}\left[(\cos\omega - \omega\sin\omega) - j(\sin\omega + \omega\cos\omega)\right]$$

by setting the imaginary part of $G(j\omega)$ equal to zero, we obtain

$$\sin\omega + \omega\cos\omega = 0$$

or

$$\omega = -\tan\omega$$

Solving this equation for the smallest positive value of ω, we obtain

$$\omega = 2.029$$

Substituting $\omega = 2.029$ into $G(j\omega)$ yields

$$G(j2.029) = \frac{K}{1 + 2.029^2}(\cos 2.029 - 2.029 \sin 2.029)$$

$$= -0.4421\ K$$

The critical value of K for stability can be obtained by letting $G(j2.029) = -1$, or

$$0.4421\ K = 1$$

Thus, the range of gain K for stability is

$$2.262 > K > 0$$

B-11-15.

$$G(s) = \frac{K}{s(s^2 + s + 4)} = \frac{0.25K}{s(0.25s^2 + 0.25s + 1)}$$

The quadratic term in the denominator has the undamped natural frequency of 2 rad/s and the damping ratio of 0.25. Define the frequency corresponding to the angle of -130° to be ω_1.

$$\angle G(j\omega_1) = -\angle j\omega_1 - \angle 1 - 0.25\omega_1^2 + j0.25\omega_1$$

$$= -90^\circ - \tan^{-1}\frac{0.25\omega_1}{1 - 0.25\omega_1^2} = -130^\circ$$

Solving this last equation for ω_1, we find $\omega_1 = 1.491$. Thus, the phase angle becomes equal to -130° at $\omega = 1.491$ rad/s. At this frequency, the magnitude must be unity, or $|G(j\omega_1)| = 1$. The required gain K can be determined from

$$|G(j1.491)| = \left|\frac{0.25K}{(j1.491)(-0.555 + j0.3725 + 1)}\right| = 0.2890K = 1$$

from which we get

$$K = 3.46$$

Note that the phase crossover frequency is at $\omega = 2$ rad/s, since

$$\angle G(j2) = -\angle j2 - \angle -0.25 \times 2^2 + 0.25 \times j2 + 1 = -90^\circ - 90^\circ = -180^\circ$$

The magnitude $|G(j2)|$ with K = 3.46 becomes

$$|G(j2)| = \left|\frac{0.865}{(j2)(-1 + j0.5 + 1)}\right| = 0.865 = -1.26 \text{ dB}$$

Thus, the gain margin is 1.26 dB. The Bode diagram of $G(j\omega)$ with k = 3.46 is shown on next page.

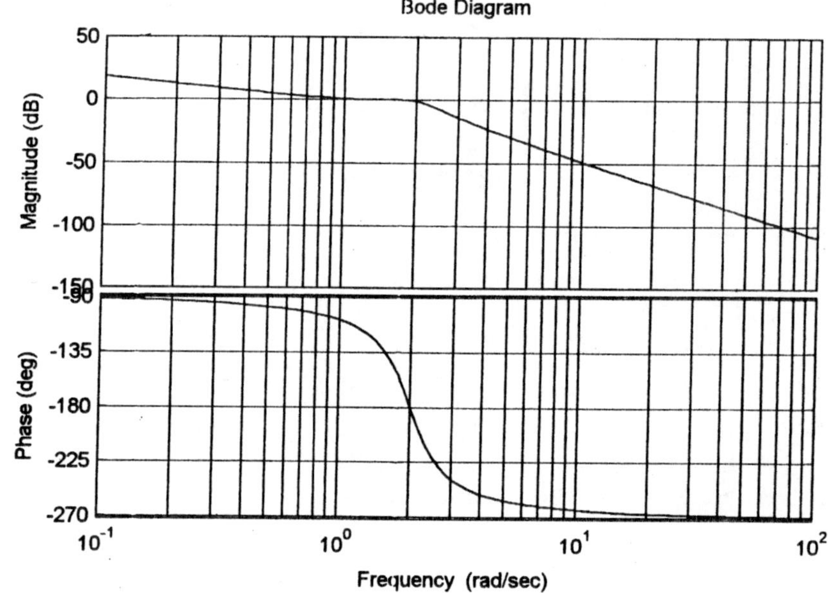

<u>B-11-16</u>. **Note that**

$$K \frac{j\omega + 0.1}{j\omega + 0.5} \frac{10}{j\omega(j\omega + 1)} = \frac{2K(10j\omega + 1)}{j\omega(2j\omega + 1)(j\omega + 1)}$$

We shall plot the Bode diagram when 2K = 1. That is, we plot the Bode diagram of

$$G(j\omega) = \frac{10j\omega + 1}{j\omega(2j\omega + 1)(j\omega + 1)}$$

The diagram is shown below. The phase curve shows that the phase angle

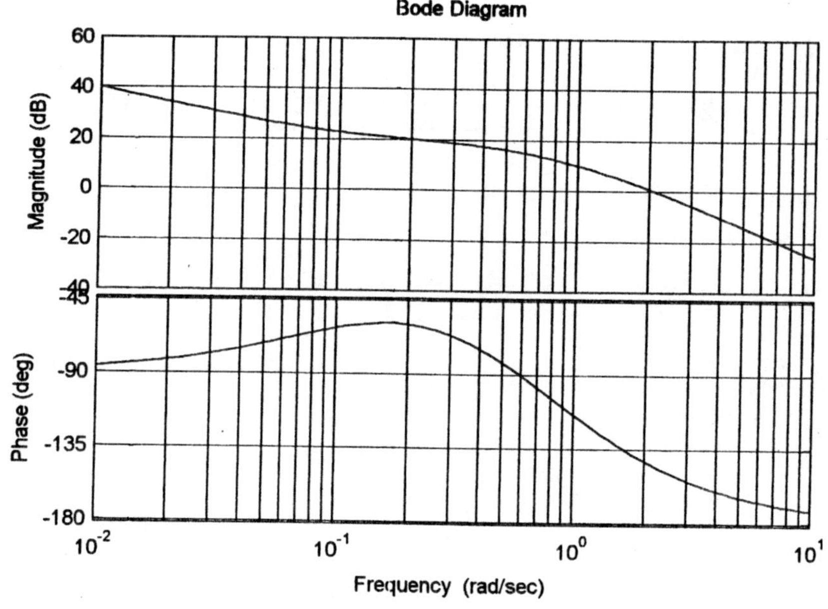

is -130° at $\omega = 1.438$ rad/s. Since we require the phase margin to be 50°, the magnitude of G(j1.438) must be equal to 1 or 0 dB. Since the Bode diagram indicates that $|G(j1.438)|$ is 5.48 dB, we need to choose 2K = -5.48 dB, or

$$K = 0.266$$

Since the phase curve lies above the -180° line for all ω, the gain margin is $+\infty$ dB.

B-11-17. Let us use the following lead compensator:

$$G_c(s) = K_c\alpha \frac{Ts + 1}{\alpha Ts + 1} = K_c \frac{s + \frac{1}{T}}{s + \frac{1}{\alpha T}}$$

Since K_v is specified as 4.0 s^{-1}, we have

$$K_v = \lim_{s \to 0} sK_c\alpha \frac{Ts + 1}{\alpha Ts + 1} \frac{K}{s(0.1s + 1)(s + 1)} = K_c\alpha K = 4$$

Let us set K = 1 and define $K_c\alpha = \hat{K}$. Then

$$\hat{K} = 4$$

Next, plot a Bode diagram of

$$\frac{4}{s(0.1s + 1)(s + 1)} = \frac{4}{0.1s^3 + 1.1s^2 + s}$$

The following MATLAB program produces the Bode diagram shown on next page.

```
>> num = [0    0    0    4];
>> den = [0.1    1.1    1    0];
>> bode(num,den);
>> grid
>> title('Bode Diagram of G(s)  = 4/[s(0.1s + 1)(s + 1)]')
```

From this plot, the phase and gain margins are 17° and 8.7 dB, respectively.

Since the specifications call for a phase margin of 45°, let us choose

$$\phi_m = 45^\circ - 17^\circ + 12^\circ = 40^\circ$$

(This means that 12° has been added to compensate for the shift in the gain crossover frequency.) The maximum phase lead is 40°. Since

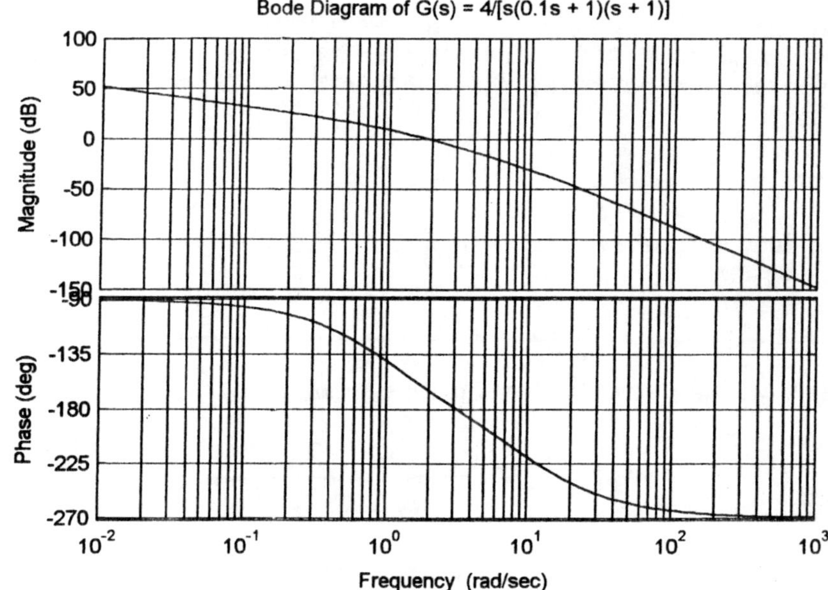

$$\sin \phi_m = \frac{1 - \alpha}{1 + \alpha} \qquad (\phi_m = 40^\circ)$$

α is determined as 0.2174. Instead of $\alpha = 0.2174$, let us choose α to be 0.21, or

$$\alpha = 0.21$$

Next step is to determine the corner frequencies $\omega = 1/T$ and $\omega = 1/(\alpha T)$ of the lead compensator. Note that the maximum phase-lead angle ϕ_m occurs at the geometric mean of the two corner frequencies, or $\omega = 1/(\sqrt{\alpha} T)$. The amount of the modification in the magnitude curve at $\omega = 1/(\sqrt{\alpha} T)$ due to the inclusion of the term $(Ts + 1)/(\alpha Ts + 1)$ is

$$\left| \frac{1 + j\omega T}{1 + j\omega\alpha T} \right|_{\omega = \frac{1}{\sqrt{\alpha} T}} = \frac{1}{\sqrt{\alpha}}$$

Note that

$$\frac{1}{\sqrt{\alpha}} = \frac{1}{\sqrt{0.21}} = 2.1822 = 6.7778 \text{ dB}$$

We need to find the frequency point where, when the lead compensator is added, the total magnitude becomes 0 dB. The magnitude $|G(j\omega)|$ is -6.7778 dB corresponds to $\omega = 2.81$ rad/s. We select this frequency to be the new gain crossover frequency ω_c. Then we obtain

$$\frac{1}{T} = \sqrt{\alpha}\,\omega_c = \sqrt{0.21} \times 2.81 = 1.2877$$

$$\frac{1}{\alpha T} = \frac{\omega_c}{\sqrt{\alpha}} = \frac{2.81}{\sqrt{0.21}} = 6.1319$$

Hence

$$G_c(s) = K_c \frac{s + 1.2877}{s + 6.1319}$$

and

$$K_c = \frac{\hat{K}}{\alpha} = \frac{4}{0.21}$$

Thus

$$G_c(s) = \frac{4}{0.21} \frac{s + 1.2877}{s + 6.1319} = 4 \frac{0.7766s + 1}{0.16308s + 1}$$

The open-loop transfer function becomes as

$$G_c(s)G(s) = 4 \frac{0.7766s + 1}{0.16308s + 1} \frac{1}{s(0.1s + 1)(s + 1)}$$

The closed-loop transfer function is

$$\frac{C(s)}{R(s)} = \frac{3.1064s + 4}{0.01631s^4 + 0.2794s^3 + 1.2631s^2 + 4.1064s + 4}$$

The following MATLAB program produces the unit-step response curve as shown below.

```
>> num = [0     0     0     3.1064     4];
>> den = [0.01631     0.2794     1.2631     4.1064     4];
>> t = 0:0.01:4;
>> c = step(num,den,t);
>> plot(t,c)
>> grid
>> title('Unit-Step Response of Compensated System')
>> xlabel('t (sec)')
>> ylabel('Output c(t)')
```

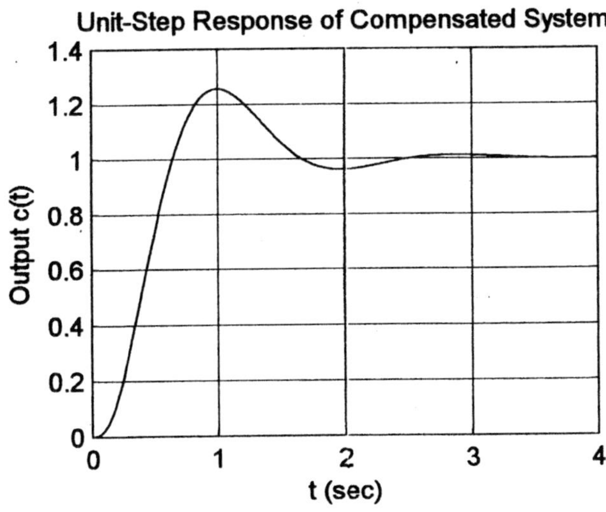

Similarly, the following MATLAB program produces the unit-ramp response curve as shown below the MATLAB program.

```
>> num = [0     0     0     0    3.1064     4];
>> den = [0.01631    0.2794    1.2631    4.1064    4    0];
>> t = 0:0.01:6;
>> c = step(num,den,t);
>> plot(t,c,t,t)
>> grid
>> title('Unit-Ramp Response of Compensated System')
>> xlabel('t (sec)')
>> ylabel('Input and Output c(t)')
```

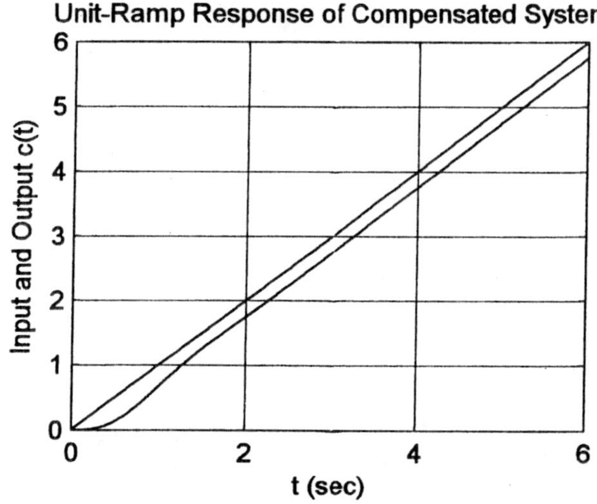

B-11-18. To satisfy the requirements, try a lead compensator $G_c(s)$ of the form

$$G_c(s) = K_c \alpha \frac{Ts + 1}{\alpha Ts + 1} = K_c \frac{s + \frac{1}{T}}{s + \frac{1}{\alpha T}}$$

Define

$$G_1(s) = KG(s) = \frac{K}{s(s + 1)}$$

where $K = K_c \alpha$. Since the static velocity error constant K_v is given as 50 s^{-1}, we have

$$K_v = \lim_{s \to 0} sG_c(s)G(s) = \lim_{s \to 0} s \frac{Ts + 1}{\alpha Ts + 1} \frac{K}{s(s + 1)} = K = 50$$

We shall now plot a Bode diagram of

$$G_1(s) = \frac{50}{s(s + 1)}$$

The following MATLAB program produces the Bode diagram shown below the program.

```
>> num = [0   0   50];
>> den = [1   1   0];
>> w = logspace(-1,2,100);
>> bode(num,den,w);
>> grid
>> title('Bode Diagram of G_1(s)')
```

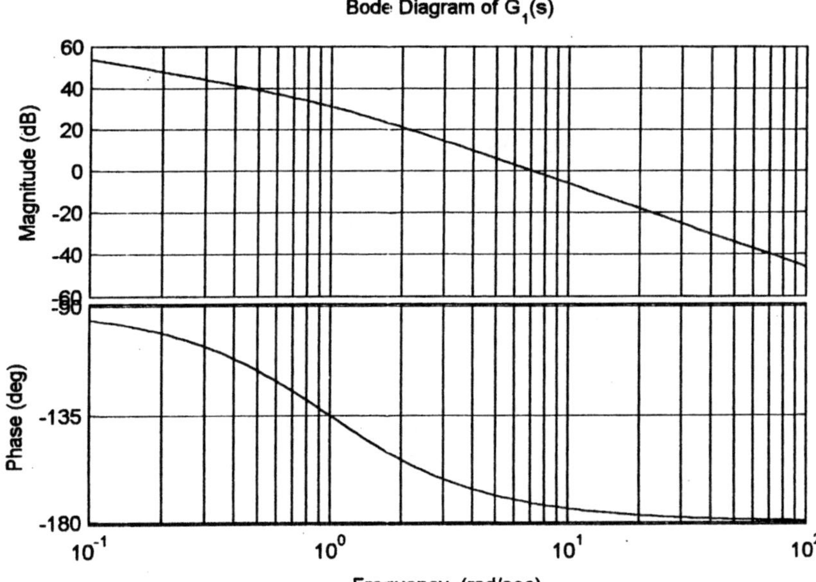

From this plot, the phase margin is found to be 7.8°. The gain margin is $+\infty$ dB. Since the specifications call for a phase margin of 50°, the additional phase lead angle necessary to satisfy the phase margin requirement is 42.2°. We may assume the maximum phase lead required to be 48°. This means that 5.8° has been added to compensate for the shift in the gain crossover frequency. Since

$$\sin \phi_m = \frac{1 - \alpha}{1 + \alpha}$$

$\phi_m = 48°$ corresponds to $\alpha = 0.14735$. (Note that $\alpha = 0.15$ corresponds to $\phi_m = 47.657°$.) Whether we choose $\phi_m = 48°$ or $\phi_m = 47.657°$ does not make much difference in the final solution. Hence, we choose $\alpha = 0.15$.

The next step is to determine the corner frequencies $\omega = 1/T$ and $\omega = 1/(\alpha T)$ of the lead compensator. Note that the maximum phase lead angle ϕ_m occurs at the geometric mean of the two corner frequencies, or $\omega = 1/(\sqrt{\alpha}T)$. The amount of the modification in the magnitude curve at $\omega = 1/(\sqrt{\alpha}T)$ due to the inclusion of the term $(Ts + 1)/(\alpha Ts + 1)$ is

$$\left| \frac{1 + j\omega T}{1 + j\omega\alpha T} \right|_{\omega = \frac{1}{\sqrt{\alpha} T}} = \left| \frac{1 + j\frac{1}{\sqrt{\alpha}}}{1 + j\alpha\frac{1}{\sqrt{\alpha}}} \right| = \frac{1}{\sqrt{\alpha}}$$

Note that

$$\frac{1}{\sqrt{\alpha}} = \frac{1}{\sqrt{0.15}} = 2.5820 = 8.239 \text{ dB}$$

We need to find the frequency point where, when the lead compensator is added, the total magnitude becomes 0 dB. The frequency at which the magnitude of $G_1(j\omega)$ is equal to -8.239 dB occurs between $\omega = 10$ and 100 rad/s. From the Bode diagram we find that the frequency point where $|G_1(j\omega)| = -8.239$ dB occurs at $\omega = 11.4$ rad/s. Noting that this frequency corresponds to $1/(\sqrt{\alpha}\,T)$, or

$$\omega_c = \frac{1}{\sqrt{\alpha}\,T}$$

we obtain

$$\frac{1}{T} = \omega_c\sqrt{\alpha} = 11.4\sqrt{0.15} = 4.4152$$

$$\frac{1}{\alpha T} = \frac{\omega_c}{\sqrt{\alpha}} = \frac{11.4}{\sqrt{0.15}} = 29.4347$$

The lead compensator thus determined is

$$G_c(s) = K_c \frac{s + \frac{1}{T}}{s + \frac{1}{\alpha T}} = K_c \frac{s + 4.4152}{s + 29.4347}$$

where K_c is determined as

$$K_c = \frac{K}{\alpha} = \frac{50}{0.15} = \frac{1000}{3}$$

Thus,

$$G_c(s) = \frac{1000}{3}\,\frac{s + 4.4152}{s + 29.4347}$$

$$= 50\,\frac{0.2265s + 1}{0.03397s + 1}$$

The following MATLAB program produces the Bode diagram of the lead compensator just designed. It is shown on next page.

```
>> num = [11.325    50];
>> den = [0.03397    1];
>> w = logspace(-1,3,100);
>> bode(num,den,w);
>> grid
>> title('Bode Diagram of G_c(s) = 50(0.2265s + 1)/(0.03397s + 1)')
```

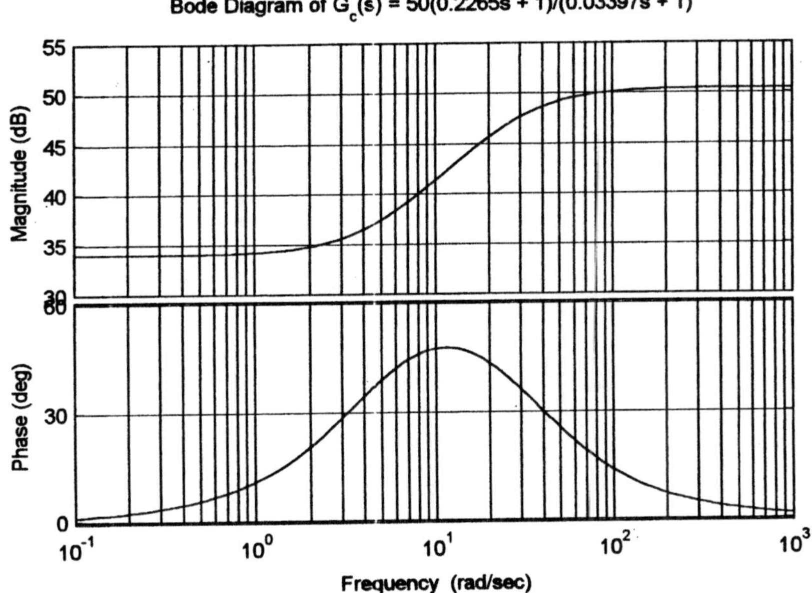

The open-loop transfer function of the designed system is

$$G_c(s)G(s) = \frac{1000}{3}\left(\frac{s + 4.4152}{s + 29.4347}\right)\frac{1}{s(s + 1)}$$

The following MATLAB program produces the Bode diagram of $G_c(s)G(s)$ which is shown on next page.

```
>> num = [0    0    1000    4415.2];
>> den = [3    91.3041    88.3041    0];
>> w = logspace(-1,3,100);
>> bode(num,den,w);
>> grid
>> title('Bode Diagram of G_c(s)G(s) = 1000(s + 4.4152)/[3(s + 29.4347)s(s + 1)]')
```

From this Bode diagram, it is clearly seen that the phase margin is approximately 52°, the gain margin is $+\infty$ dB, and $K_v = 50\ s^{-1}$; all specifications are met. Thus, the designed system is satisfactory.

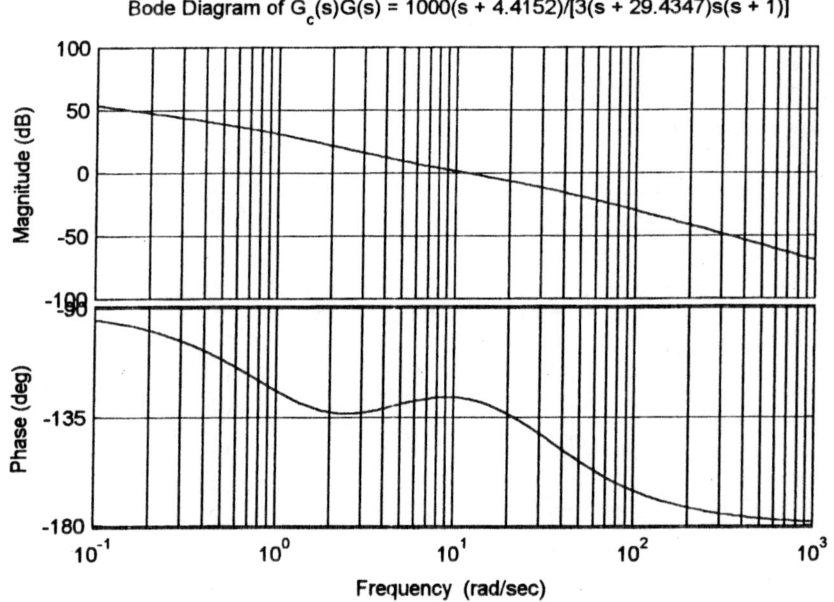

Next, we shall obtain the unit-step and unit-ramp responses of the original uncompensated system and the compensated system. The original uncompensated system has the following closed-loop transfer function:

$$\frac{C(s)}{R(s)} = \frac{1}{s^2 + s + 1}$$

The closed-loop poles of the uncompensated system are as follows:

$$s = -0.5 \pm j0.866$$

The closed-loop transfer function of the compensated system is

$$\frac{C(s)}{R(s)} = \frac{1000(s + 4.4152)}{3(s + 29.4347)s(s + 1) + 1000(s + 4.4152)}$$

$$= \frac{1000s + 4415.2}{3s^3 + 91.3041s^2 + 1088.3041s + 4415.2}$$

The closed-loop poles of the compensated system are as follows:

$$s = -11.1772 \pm j7.5636, \qquad s = -8.0804$$

The MATLAB program given below produces the unit-step responses of the uncompensated and compensated systems. The resulting response curves are shown below the MATLAB program.

```
>> num = [0    0    1];
>> den = [1    1    1];
>> numc = [0    0    1000    4415.2];
>> denc = [3    91.3041    1088.3041    4415.2];
>> t = 0:0.01:8;
>> c1 = step(num,den,t);
>> c2 = step(numc,denc,t);
>> plot(t,c1,t,c2)
>> grid
>> title('Unit-Step Response of Uncompensated and Compensated Systems')
>> xlabel('t (sec)')
>> ylabel('Outputs')
>> text(1,1.25,'Compensated system')
>> text(1.7,0.25,'Uncompensated system')
```

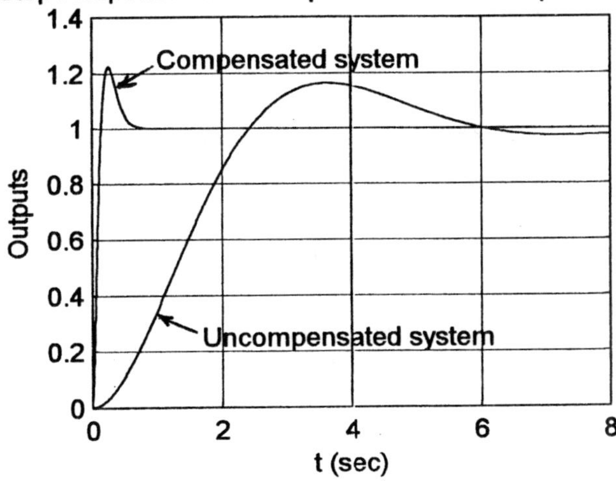

The MATLAB program given on next page produces the unit-ramp responses of the uncompensated system and compensated system. The response curves are shown on next page.

From the unit-ramp response curves it can be seen that the output of the compensated system follows the input ramp very closely. For the compensated system, the error in following the ramp input can be seen for $0 < t < 0.5$, but it is almost zero for $0.5 < t$. (The steady-state error in the unit-ramp response is 0.02.)

```
>> num = [0    0    0    1];
>> den = [1    1    1    0];
>> numc = [0    0    0    1000    4415.2];
>> denc = [3    91.3041    1088.3041    4415.2    0];
>> t = 0:0.01:8;
>> c1 = step(num,den,t);
>> c2 = step(numc,denc,t);
>> plot(t,c1,t,c2,t,t)
>> title('Unit-Ramp Response of Uncompensated and Compensated Systems')
>> xlabel('t (sec)')
>> ylabel('Input and Outputs')
>> text(1,7,'Compensated system')
>> text(3.5,1.7,'Uncompensated')
>> text(3.5,1,'system')
>> text(1,4,'Input')
```

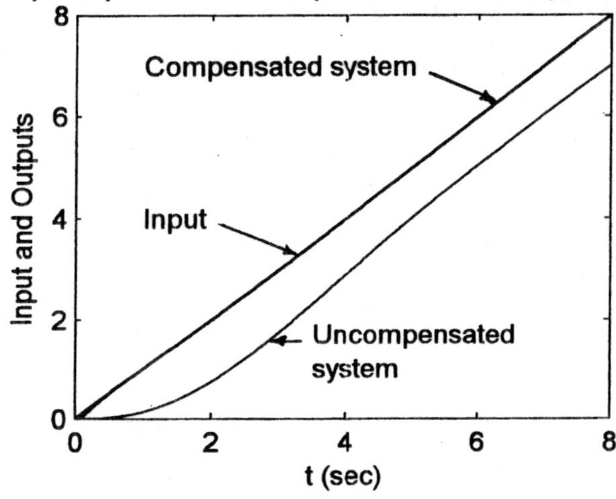